Thermophiles for Biotech Industry

Jujjavarapu Satya Eswari • Swasti Dhagat
Ramkrishna Sen

Thermophiles for Biotech Industry

A Bioprocess Technology Perspective

Springer

Jujjavarapu Satya Eswari
Department of Biotechnology
National Institute of Technology
Raipur, Chhattisgarh, India

Swasti Dhagat
Department of Biotechnology
National Institute of Technology
Raipur, Chhattisgarh, India

Ramkrishna Sen
Department of Biotechnology
Indian Institute of Technology
Kharagpur, West Bengal, India

ISBN 978-981-32-9918-4 ISBN 978-981-32-9919-1 (eBook)
https://doi.org/10.1007/978-981-32-9919-1

This Springer imprint is published by the registered company Springer Nature Singapore Pte Ltd.
The registered company address is: 152 Beach Road, #21-01/04 Gateway East, Singapore 189721, Singapore

Contents

About the Authors

Jujjavarapu Satya Eswari is currently working as an Assistant Professor in the Department of Biotechnology at the National Institute of Technology (NIT), Raipur, India. She completed her Ph.D. from IIT Hyderabad/IICT Hyderabad (CSIR lab). She also pursued her research in the areas of bioinformatics and bioprocess and product development. She gained pioneering expertise in the application of mathematical and engineering tools in biotechnological processes. Her fields of specializations include bioinformatics, biotechnology, process modeling, evolutionary optimization, and artificial intelligence. She has more than 35 publications in SCI/Scopus indexed journals and 35 proceedings in international and national conferences. Her research contributions have received wide global citations. She has also published six book chapters and four books (currently in press) with international publishers. Dr. Jujjavarapu Satya Eswari completed two sponsored and currently one ongoing project. She has a teaching experience of more than 7 years along with a research experience of 3 years. She has conducted various short-term courses on bioprocess development and soft computing and intelligent techniques. She has delivered invited lectures in various institutes. She visited three countries for international conferences. She has guided more than 47 B.Tech. and postgraduate students and is currently supervising 5 Ph.D. students, a project assistant, and 7 B.Tech. students. She is an active participant in both academic and research activities. She has reviewed various research articles in 20 international journals. She is an active member of various organizations. She has received Early Career Research (ECR) Award by the DST, IEI Young Engineer Award, Venus Outstanding Woman in Engineering, and DK International Best Faculty Award (2018).

Swasti Dhagat is currently a Research Scholar in the Department of Biotechnology at the National Institute of Biotechnology. She has completed her M.Tech. in Biotechnology from Birla Institute of Technology and Science (BITS), Pilani, Hyderabad, and B.Tech. in Biotechnology from Banasthali University. She has seven publications in international peer-reviewed journals, few conference proceedings in international and national conferences, and a book which is currently in press. She has pursued research from the Council of Scientific and Industrial Research (CSIR) laboratories in India in the areas of microbiology and environmental biotechnology.

Ramkrishna Sen is a Professor of Bioprocess Engineering and currently Head of the Department of Biotechnology and a Joint Faculty, P K Sinha Center for Bioenergy & Renewables, IIT Kharagpur. He served as a Fulbright Visiting Professor in Columbia University, New York, USA. He also worked as an Assistant Professor in BITS, Pilani, for about 2 years and as Manager (R&D Biotech) in Cadila Pharmaceuticals Ltd., Ahmedabad, for about 3 years. He has been engaged in R&D activities in the areas of healthcare, energy, environment, and water, with a focus on green process and product development in microalgal-microbial biorefinery models. So far, 21 students have completed their Ph.D. under his guidance. Currently, 18 Ph.D. students including 2 in foreign universities are working under his supervision. He has about 230 publications in international and national journals, books, and conference proceedings and 13 patent applications including 3 international applications to his credit. Prof. Sen completed 24 sponsored and consultancy projects and has 9 ongoing projects. He published a book titled *Biosurfactants* (Springer, USA). Prof. Sen has been honored with the Distinguished Alumni Award (Jadavpur University, 2017), Fulbright Visiting Professor Fellowship (USA, 2013–2014), UKIERI Award for Academic Visits (UK, 2007), and National Award for Technology Innovation, Government of India (2012). He is a Life Member of Indian Institute of Chemical Engineers (IIChE) and Member of European and Asian Federations of Biotechnology. Prof. Sen administered as Vice-Chairman and Chairman (GATE & JAM), IIT Kharagpur, from 2015 to 2018.

Introduction 1

1 Diversity of Archaebacteria

Microorganisms are organisms that are too small to be seen by the naked eyes. They may be single-celled (unicellular) or multicellular. Microorganisms are widely present in nature. They include bacteria, archaea, fungi, protozoa, algae, and viruses. Archaebacteria or archaea are unicellular prokaryotes. They differ from bacteria in their structure of DNA which resembles the DNA of eukaryotic cells and their cell wall structure which lack peptidoglycan (Table 1.1). Archaeans have been known to exist in extreme environmental conditions which are not suitable for most life forms. Because of their existence in extreme conditions, they are also called as extremophiles. They are called as living fossils as they are thought to be the first life forms to exist on Earth by absorbing solar radiations under anaerobic conditions.

Based on their habitat they are classified as:

- Halophiles

Halophiles live in extremely salty conditions. Their habitat includes salty soil and water-like salt marshes, salt pans, and salt beds where the salt concentration can be as high as 2.5–5.0 M. Examples of halophiles are *Halococcus* and *Halobacterium*. Halophiles are aerobic chemoheterotrophs. The red carotenoid pigment present in their cell membranes protects these archaebacteria from harmful solar radiations. Halophiles cannot utilize external materials under anaerobic conditions, and hence under these conditions, they survive on ATP synthesized by the carotenoid pigment system. Halophiles live in extreme salt conditions because of the presence of special lipids in their cell membranes, mucilage covering, high internal salt content, and absence of sap vacuoles leading to plasmolysis.

J. S. Eswari et al., *Thermophiles for Biotech Industry*,
https://doi.org/10.1007/978-981-32-9919-1_1

Table 1.1 Comparison between bacteria, archaea, and eukaryotes

Characteristics	Bacteria	Archaea	Eukaryotes
Peptidoglycan in cell wall	Present	Absent	Absent
Fatty acid in membrane lipids	Fatty acids linked by ester bonds to glycerol	Fatty acids linked by ether bonds to glycerol	Fatty acids linked by ester bonds to glycerol
RNA polymerase	Consists of 4 polypeptides	Consists of more than 8 polypeptides	Consists of 10–12 polypeptides
Synthesis of proteins	Initiator tRNA with modified methionine	Initiator tRNA with unmodified methionine	Initiator tRNA with unmodified methionine
Metabolism	Nitrogen fixation, denitrification, chemolithotrophy	Nitrogen fixation, denitrification, chemolithotrophy, methanogenesis	Not present

- Psychrophiles

Psychrophiles are cold-loving bacteria which can thrive at a temperature of 0 °C or lower. Psychrotrophs, on the other hand, are cold-tolerant bacteria that can grow at 0 °C or lower. Their optimal growth temperature is 15 °C with a maximal growth temperature of 20 °C. Their ability to grow at low temperatures can be correlated to the increased fluidity of cell membrane which is because of the increase in the amount of unsaturated fatty acids in the lipid phase of cell membrane and a protein conformation which is functional at low temperatures. They are widely spread in natural environment and as a contaminant in foods. Some of them are also pathogenic to man, animals, and plants. They can play a major role in biodegradation of organic matter during cold seasons. It is considered that the ability of psychrophiles to survive under low-temperature conditions is because of the presence of one or more thermolabile cellular components, especially enzymes. Enhancement of some degradative activities results in exhaustion of cell energy and leakage of intracellular substances leading to cell lysis (Gounot 1986; Moyer and Morita 2007).

- Thermophiles

Thermophiles dwell in hot environments. They have the ability to tolerate high temperatures and acidity. They survive in acidic and hot regions such as in hot springs where temperature can reach as high as 80 °C. Some of the examples of thermophiles belong to the genus *Thermoplasma* and *Thermoproteus*. It is believed that the thermophiles die when they are exposed to temperatures below 55 °C. They are chemosynthetic organisms, that is, they obtain energy for synthesis of food by oxidizing sulfur. They oxidize sulfur to sulfuric acid under aerobic conditions as per the following reaction:

$$2S + 2H_2O + 3O_2 \rightarrow 2H_2SO_4$$

Under anaerobic conditions, they might reduce sulfur to hydrogen sulfide. Precipitation of bicarbonates into carbonate form occurs as a result of this reaction. The ability of thermophiles to tolerate high temperature and acidity is because of the presence of resistant enzymes that can operate under acidic conditions and branched chain lipids in cell membranes.

- Methanogens—methane-producing organisms

Methanogens are strict anaerobes, that is, they grow without oxygen. They are autotrophs, and they obtain carbon and energy from decomposition of products. They are found in marshes, swamps, and intestinal tract of animals and some humans where they convert formic acid and carbon dioxide to methane in the presence of hydrogen. Some examples of methanogens are *Methanobacterium* and *Methanococcus*. This ability is exploited for the production of methane and fuel gas in biogas plants; *Methanobacterium* live as symbiont inside rumen (first chamber of stomach) of herbivores that chew their cud. These bacteria help these ruminants in fermentation of cellulose.

The physiology of archaeans is the key for their survival in extreme conditions. The various enzymes synthesized by them protect their cells from the destructive effects of acidic or salty water and high temperatures. Carbon dioxide, hydrogen gas, and sulfur are the different energy sources of archaebacteria. They also utilize sunlight for energy production with a mechanism that is different from plants. Archaea absorb sunlight using bacteriorhodopsin, the membrane pigment of archaea. This membrane pigment reacts with light which leads to the formation of adenosine triphosphate (ATP), the energy molecule. Archaebacteria are being utilized for the production of thermostable enzymes, antibiotics, biogas, and organic solvents and in sewage treatment. They are also being used in the fermentation of cellulose.

2 Thermophilic Microorganisms

Thermophilic microorganisms survive at higher temperatures ranging from 41 to 122 °C. They are included in the domain archaea. They can be classified into:

- Obligate thermophiles which require high temperature for growth
- Facultative thermophiles which can survive at higher temperature and also at temperatures below 50 °C
- Hyperthermophiles which require temperatures above 80 °C

Heat-loving thermophiles are found in many adverse environments found on Earth. They can be found in hot springs or hydrothermal vents in ocean. They can survive in minerals, gases, or metals that are found in extreme conditions. Some of the examples of thermophilic microorganisms are *Pyrolobus fumarii*, *Chloroflexus aurantiacus*, *Thermus aquaticus*, *Thermus thermophilus*. A thermophile *Methanopyrus kandleri* holds the record for existing in hottest temperature and has found to survive at around 120 °C.

The difference in the physiological characteristics of bacterial and archaeal lipids, particularly the increased proportion of certain lipids as the growth temperature of bacteria increases, the importance of chemical stability of lipids in heat-loving microorganisms, and the properties of lipid bilayer in cell membrane which allow the membranes to work at high temperatures enable the thermophilic microorganisms to grow and survive at high temperatures. The adaptation of thermophiles to survive at high temperatures is not because of one chemically stable lipid. The lipids in cell membrane need to be in a liquid crystalline state and require high permeability barrier.

The survival of thermophilic bacteria at high boiling temperatures as compared to mesophiles is possible because of their thermostable cell components. The temperature range of growth of sulfur-utilizing archaebacteria is approximately 30 °C. This is because of some of the properties of cell, for example, the conformation of nucleic acids and the fluidity of cell membrane. This phenomenon might also be responsible for their absence of growth below 60–82 °C. Alternatively, these microorganisms can survive at low temperatures for years (Stetter et al. 1983). The low molecular weight compounds, such as NAD and ATP, undergo rapid hydrolysis at 100 °C, whereas thermolabile amino acids get decomposed (Bernhardt et al. 1984). The microorganisms can survive at these temperatures by resynthesizing these components to gain biological stability to heat and hence require the presence of sufficient substrates. This is demonstrated by the facts that in the absence of substrates, *Pyrodictium* gets killed rapidly at 110 °C, and the optimal temperature for growth of *Staphylococcus marinus* is 7 °C lower in a minimal medium than the full medium (Fiala et al. 1986). Even though the upper temperature for microbial growth is not defined, it might be possible that it is under 150 °C where heat-labile molecules are resynthesized successfully (Stetter et al. 1986a). To summarize, the survival of thermophiles at higher temperatures is because of the following factors:

1. Thermophilic enzymes and proteins exhibit some differences in their amino acid composition from the homologous mesophilic enzymes and proteins. Amino acid substitution in thermophilic enzymes at few critical locations results in a protein folding that is more heat stable as compared to the mesophilic counterparts, and hence they are functional at such high temperatures.
2. The presence of saturated fatty acid-rich lipids in their cytoplasmic membrane provides them stability and functionality at higher temperatures. A hydrophobic environment is created by saturated fatty acids as compared to unsaturated fatty acids which help in maintaining thermophilic membrane stability.
3. Heat stability of thermophilic proteins can be increased by increasing the number of ionic bonds which is formed between the positive and negative charges of amino acids. The interaction results in dense packaging of highly hydrophobic interiors of the proteins which protects the proteins from unfolding in aqueous cytoplasm. Thermophilic proteins can be stabilized against thermal degradation by the production of high amounts of solutes, namely, diglycerol phosphate, inositol phosphate, and manosylglycerate in the cytoplasm of some thermophiles.

4. The membranes of thermophilic and hyperthermophilic archaebacteria do not contain fatty acids. Instead, they have lipids with branched hydrocarbon chains which are made up of repeats of five carbon compounds called as isoprene. These units are bonded by ether linkages. These linkages stabilize the membranes against thermal damage, and the branched hydrocarbon chain decreases membrane fluidity. Hence, the membranes of thermophiles are stabilized under high temperatures. The formation of lipid monolayer of bacterial cytoplasmic membrane provides it additional thermostability than the lipid bilayer present in many species of bacteria and eukaryotes.

3 Extremophilic Fungi

Fungi include yeasts, mushroom, molds, mildews, rusts, and smuts. These are classified as eukaryotes, that is, they have a true nucleus. They are heterotrophs, that is, they cannot synthesize their own food but depend on organic substances, mainly animal or plant matter to derive nutrients. Thermophilic fungi grow at temperatures above 20 °C with a limited growth at temperature of 60–62 °C. The optimal temperature required by thermophilic fungi for their growth is 45 °C. It has been demonstrated that eukaryotes are incapable to grow at higher temperatures as they cannot form organellar membranes with active functionality and thermostability. Hence the growth of thermophilic fungi was restricted to 60 °C (Tansey and Brock 1972). Fungi have a wide variety of applications. Apart from being used as a food source, in food industry, fungi are also used in ripening of cheese and fermentation of grains and fruits to produce beer and wine. Yeast, *Saccharomyces cerevisiae*, is an important component in bread and bakery products. Because of their pathogenicity and specificity to the insect they attack, they can be used as potential pesticides and insecticides. Symbiotic associations of fungi with plants help in the growth and survival of many plant species. As secondary metabolites, fungi produce many antibiotic compounds, for example, penicillin. They also produce precursors of steroid hormones, cyclosporine, and ergot alkaloids. They are also used as model research organism for eukaryotes. Other than these applications, thermophilic fungi are applied for the production of many thermostable enzymes. The extracellular or secretory enzymes from thermophilic fungal species are protease, lipase, α-amylase, α-D-glucuronidase, cellobiose dehydrogenase, cellulase, D-glucosyltransferase, glucoamylase, laccase, phytase, polygalacturonase, and xylanase. The cell-associated enzymes are ATP sulfurylase, β-glucosidase, the enzymes belonging to pentose phosphate pathway and TCA cycle, invertase, lipoamide dehydrogenase, protein disulfide isomerase, and trehalase (Maheshwari et al. 2000). Table 1.2 summarizes some of the potential applications of thermophilic fungi along with their optimal growth temperature. The extracellular enzymes provide the advantages of lesser downstream processing steps as it does not require cell disruption and removal of cellular debris. Figure 1.1 shows a simple process layout of extracellular enzyme production.

Table 1.2 List of thermophilic fungi

Microorganism	T_{growth} (°C)	Potential products	References
Canariomyces thermophile	45	Protease	Salar and Aneja (2007) and Srilakshmi et al. (2014)
Chaetomidium pingtungium	48		Salar and Aneja (2007)
Chaetomium mesopotamicum	30–52	Protease	Salar and Aneja (2007) and Srilakshmi et al. (2014)
Chaetomium senegalensis	45		Salar and Aneja (2007)
Chaetomium thermophilum	50–55	β-1,4-Xylanase, endocellulase, β-D-glucosidase, glucoamylase	Venturi et al. (2002), Hakulinen et al. (2003), Li et al. (2003), Chen et al. (2005), and Bock et al. (2014)
Corynascus thermophilus	45	Cellobiose dehydrogenase	Salar and Aneja (2007) and Tasca et al. (2011)
Coonemeria aegyptiaca	55		Ueda and Udagawa (1983)
Melanocarpus albomyces	50	Laccase, cellulose, xylanase	Maheshwari and Kamalam (1985), Jain (1995), Hakulinen et al. (2002), Kiiskinen et al. (2002), and Miettinen-Oinonen et al. (2004)
Myceliophthora thermophile	45–50	Endoglucanase, aldonolactonase, laccase	Roy et al. (1990), Beeson et al. (2011), and Lloret et al. (2012)
Papulaspora thermophila	45	Cellulolytic and amylolytic enzymes	Fergus (1971) and Chapman et al. (1975)
Rhizomucor miehei	55	Lipase	Schipper (1978) and Rodrigues and Fernandez-Lafuente (2010)
Rhizomucor nainitalensis	48	Rennet	Joshi (1982) and Khademi et al. (2013)
Rhizomucor pusillus	50	Flax retting, polygalacturonase	Schipper (1978) and Henriksson et al. (1999)
Scytalidium thermophilum	45	Endoglucanase, β-glucosidase, α-amylase, cellulase, xylanase, D-xylosidase	Straatsma et al. (1994), Ögel et al. (2001), Aquino et al. (2003), Zanoelo et al. (2004a, b), and Jatinder et al. (2006)
Talaromyces thermophilus	45–50	β-Glucosidase, xylanase, lipase, β-D-xylosidase, cellulase, glucoamylase, protease	Maheshwari et al. (2000), Nakkharat and Haltrich (2006), Guerfali et al. (2008), Romdhane et al. (2010), and Srilakshmi et al. (2014)
Thermoascus aurantiacus	45	Pectinase, xylanase, β-glucosidase	Tan et al. (1987), Souza et al. (1999), Martins et al. (2002), Salar and Aneja (2007), and Leite et al. (2008)
Thermomyces lanuginosus	52–53	Amylase, lipase, hemicellulose, phytase, xylanase, amylolytic enzymes, mannase, pectinase	Cooney and Emerson (1964), Purkarthofer et al. (1993), Berka et al. (1998), Puchart et al. (1999), Nguyen et al. (2002), Singh et al. (2003), Kunamneni et al. (2005), and Fernandez-Lafuente (2010)

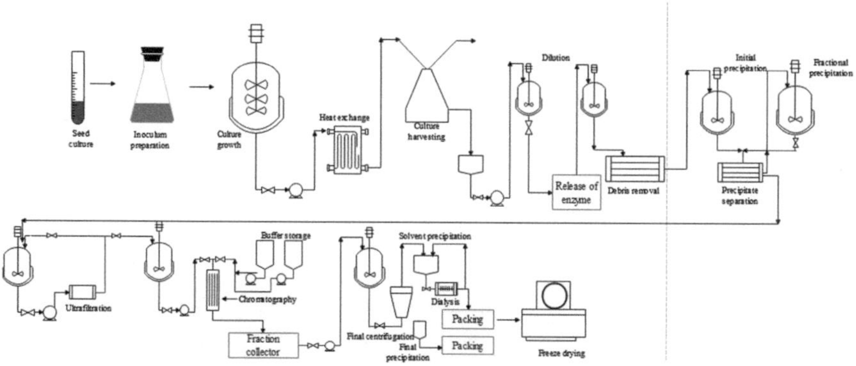

Fig. 1.1 Process layout for the production of endocellular enzymes

4 Habitat and Diversity of Hyperthermophilic Microorganisms: Challenges to the Living World

Microorganisms that grow and reproduce at very high temperatures of 60–113 °C are called hyperthermophiles. They have an optimal temperature above 80 °C, whereas some can work optimally at 100 °C. Apart from high temperatures, hyperthermophiles can also tolerate high acidity or high radiation levels. *Sulfolobus acidocaldarius* is the first acidophilic hyperthermophile to be isolated by Thomas D. Brock from acidic, hot spring of Yellowstone National Park, Wyoming, in the late 1960s. From that time more than 70 species of hyperthermophiles have been isolated. Hyperthermophiles are categorized into archaea domain though some anaerobic photosynthetic bacteria and cyanobacteria which grow at 70–75 °C are also included in this group. Most of the hyperthermophiles, namely, *Pyrolobus*, *Pyrococcus*, and *Pyrodictium*, are anaerobic. A hyperthermophile found in deep-sea vents, *Methanopyrus*, has ancient genetic composition. They are thought to be one of the oldest organisms on Earth. Another hyperthermophile, *Pyrolobus fumarii*, is a nitrate-reducing chemolithotroph and belongs to Crenarchaeota. This grows at the walls of marine Atlantic hydrothermal vents, multiplies at 105 °C, and reproduces at 113 °C. Their growth is inhibited in cooler environments of 90 °C or lower. *Pyrococcus furiosus* is found near a volcanic vent in Italy.

A hyperthermophilic archaeon strain, Strain 121, was discovered in the Pacific Ocean. It was found to thrive at 121 °C and double its population in an autoclave at 121 °C within 24 h. This strain can survive for 2 h at 130 °C but cannot reproduce until the medium is refreshed and cooled to 103 °C. The highest growth temperature of 122 °C is recorded for *Methanopyrus kandleri* which is found in Central Indian Ridge. Although the upper temperature limit of hyperthermophiles is not determined, the discovery of microorganisms surviving at 150 °C (super-hyperthermophiles) or more will be surprising as at such high temperatures the chemical bonds present between DNA and other important biomolecules will break down without any mechanism to prevent it from occurring.

Thermophiles grow rapidly at 80–106 °C temperatures. At such high temperature, the small biomolecules present within the cells are degraded. Currently, hyperthermophiles inhabit water present in hot environments. On land, these can be found in the soil and surface water near volcanic exhalations. A variety of hyperthermophiles can be found in hydrothermally heated rocks of deep-sea hot smoker chimneys.

Hyperthermophiles inhabit various environmental conditions with salinity, pH, temperature, redox potential, gases, and minerals. Their mode of nutrition is chemolithoautotrophic wherein energy source is inorganic redox reaction (chemolithotrophic) and carbon dioxide serves as the carbon source (autotrophic). These hyperthermophiles undergo both aerobic and anaerobic types of respiration. Hydrogen, sulfur, sulfide, and ferrous ions act as electron donors. Some hyperthermophilic microorganisms may even use oxygen as electron acceptor. Microaerophilic hyperthermophiles are oxygen-respiring and grow only in the presence of reduced concentrations of oxygen. Anaerobic respiration includes nitrate, sulfur, sulfate, ferric-iron, and carbon dioxide respirations. Some hyperthermophiles are facultative heterotrophs that can utilize organic matter (decaying, dead matter) instead of inorganic materials from the environment. Their mode of respiration and energy generation is aerobic, anaerobic, or by fermentation.

The small size (micrometers) of hyperthermophilic cells is inefficient to provide any protection and insulation to the cells against hot environment. Hence, the cell components must themselves be heat resistant. The molecular mechanism behind this heat resistance is currently under investigation. As hyperthermophiles belong to two distinct domains of life (bacteria and archaea), the molecular mechanisms of heat resistance and adaptation of both the domains may be different depending upon the position of organism in phylogeny.

Heat-sensitive components of the cell include lipids, proteins, and nucleic acids. These components are decomposed at the temperature in which hyperthermophiles grow and reproduce. Hence, cell components of hyperthermophiles have been modified to protect them from thermal damage. For example, the hyperthermophilic bacteria *T. maritime* contains 15,16-dimethyl-30-glyceryloxy-triacontanedioic acid in the membrane lipid of the organism. This novel glycerol ether lipid, present in hyperthermophiles but not in mesophiles, increases the stability of membranes against hydrolysis at high temperatures. The cell membranes of all archaean species are made up of ether lipids. These lipids are derived from diphytanyl glycerol or its dimer (di(biphytanyl) diglycerol) and show resistance against hydrolysis at high temperatures and low or acidic pH. Reverse gyrase, a type I DNA topoisomerase, improves thermal resistance of DNA double helix in hyperthermophiles. Reverse gyrase is also responsible for causing positive twists for stabilization. The core histones of archaeal hyperthermophiles have been found to be phylogenetically related to eukaryotic histones. The addition of histones to purified DNA in vitro drastically increases its melting point. Increase in GC base pair content in the stem areas stabilizes the secondary structure of ribonucleic acid against thermal destruction along with posttranslational modification. Enzymes purified from hyperthermophiles exhibit unusual thermal stability in vitro. Amylase isolated from *Pyrococcus woesei* shows activity even at 130 °C. The comparison between three-dimensional structures of hyperthermophilic proteins and their homologous mesophilic enzymes can assist in comprehending the stabilizing principles of hyperthermophilic proteins. Heat-shock proteins play an important role in the upper temperature range of hyperthermophiles. It has been shown that at a temperature of 108 °C, a molecular chaperone, thermosome, induced by heat, make up 80% of the soluble protein fraction of *Pyrodictium occultum* crude extract (Phipps et al. 1993). After the complete induction of thermosomes, the cultures of *P. occultum* can survive up to 1 h during autoclaving (121 °C).

The stability of biomolecules determines the upper limit of hyperthermophiles which is still unknown. As the cellular temperature reaches 100 °C, the low molecular weight cellular compounds (NAD and ATP) undergo rapid hydrolysis, and the hydrophobic interactions weaken. This temperature is also responsible for thermal decomposition of few amino acids. Hyperthermophiles grow and survive at these temperatures by re-synthetizing the thermosensitive molecules. In order to maintain electrochemical proton gradients to gain energy, the permeability of membrane to protons becomes very high at higher temperatures. It is a challenging task for microbiologists to discover and grow such exciting microbes (Stetter 1999).

More than 70 species of hyperthermophiles have been identified by the end of 1999. Till date, isolation and purification of hyperthermophilic cultures is a challenging task. The vast diversity of hyperthermophiles can be thought of by studying the 16S rRNA of samples isolated from one hot spring with analysis of environmental lipid. *Thermocrinis ruber*, a pink-filament-forming bacterium, was cultivated 25 years after this organism was described by Brock in 1967 (Brock 1967; Huber et al. 1998). In the near future, scientists will have to develop new isolation and purification techniques for novel hyperthermophilic microorganisms with different metabolic requirements and diverse metabolic products. Hyperthermophiles have been isolated from many natural environments, namely, hydrothermal vents located 4000 m below sea level, deep geothermal heated oil-containing stratifications, deep-sea and shallow marine hot sediments, and continental solfataras with temperatures ranging from 80 to 115 °C. Hyperthermophiles have also been known to survive in environmental conditions with pressure in the range of 200–360 atm. These hyperthermophiles are classified as barophile or barotolerant microorganisms.

Of the 70 species of hyperthermophiles identified till now, most of them belong to the domain archaea, except for *Thermotogales* and *Aquificales* which are bacteria. As these two bacteria lie at the bottom of the bacterial lineage, they are of interest in establishing evolutionary studies. The genome sequence of *Thermotoga maritima* suggests the lateral transfer of genes between archaea and bacteria. Almost one-fourth of *T. maritima* proteins, encoded by bacterial open reading frames, exhibit similarity to archaeal proteins than to bacterial proteins. The gene order of 81 of these gene clusters is identical to archaea. Also, approximately 80% of the *T. maritima* genes with significantly different G+C content have archaea-like genes. A 16S rRNA, isolated from a hyperthermophilic environment, suggested the presence of a third branch of archaeal domain other than Euryarchaeota and Crenarchaeota, the Korarchaeota. Korarchaeota hold a deeper position in archaeal tree than

Euryarchaeota and Crenarchaeota. Crenarchaeota includes halophiles and thermoacidophiles, whereas Euryarchaeota contains mesophilic methanogens.

Hyperthermophiles can be primary producers or decomposers of organic matter. The primary producers are chemolithotrophs and include methanogens, sulfur oxidizers, and sulfur reducers. On the basis of high sulfur content of natural environments, hyperthermophiles can be categorized into facultative or obligate chemolithotrophs. They can either reduce elemental sulfur (S^0) to hydrogen sulfide (H_2S) with hydrogen atoms (H_2) as in anaerobes or oxidize sulfur (S^0) to sulfuric acid with oxygen atoms (O_2) as seen in aerobes. Sulfolobales are extremely acidophilic hyperthermophiles and are strict aerobes (*Sulfolobus*) or facultative aerobes (*Acidianus*). These have been isolated from continental solfataras. All members of *Thermotogales* and few members of *Thermococcales* and *Pyrococcales* grow without sulfur and obtain their energy from fermentation. As hyperthermophiles have low organic matter, they acquire energy and carbon from a mixture of peptides obtained from the decomposition of producers. Only a few species of hyperthermophiles can use polysaccharides. *Archaeoglobus profundus* is the only hyperthermophilic microorganism that is known to utilize organic acids.

Extremely acidophilic hyperthermophiles thrive in acidic hot surfaces of solfataric fields which are exposed to oxygen. They have lobes and are coccoid in shape. They are aerobes and facultative aerobes and grow between pH 1 and 5 with optimum pH of 3. They are phylogenetically classified into *Acidianus*, *Metallosphaera*, and *Sulfolobus* archaeal genera. They form the domain Sulfolobales with strictly anaerobic *Stygiolobus*. Strict aerobes, *Sulfolobus*, have both autotrophic and heterotrophic growth. During their autotrophic growth, they oxidize sulfur atoms from hydrogen sulfide and sulfur dioxide to yield sulfuric acid or water. In heterotrophic growth, the microorganisms use sugars, yeast extract, and peptone as energy sources. *Acidianus* grow anaerobically or aerobically. They oxidize hydrogen and use elemental sulfur and oxygen as electron acceptors for anaerobic and aerobic oxidation, respectively.

The strict anaerobic hyperthermophiles that live in terrestrial hot springs with low salinity are slightly acidophilic or neutrophilic and belong to *Desulfurococcus*, *Pyrobaculum*, *Sulfophobococcus*, *Thermofilum*, *Thermosphaera*, and *Thermoproteus* genera. *Pyrobaculum*, *Thermofilum*, and *Thermoproteus* cells are rod-shaped, stiff cells. *Desulfurococcus* are coccoid-shaped cells which are heterotrophic sulfur respirers. *Sulfophobococcus* and *Thermosphaera* are coccoid microorganisms that undergo fermentation. *Methanothermus* are oxygen-sensitive chemolithoautotrophs and grow in the temperature range of 65–97 °C. They obtain energy by reducing carbon dioxide by hydrogen (Vieille and Zeikus 2001).

5 Applications of Thermophiles

The heat-loving organisms can be exploited for their enzymes, called as extremozymes, and molecular chaperones. These products can be used in various biotechnological and industrial processes.

5.1 Extremozymes

Extremozymes are the enzymes that are functional at extreme environmental conditions, namely, high temperature and low pH. As the industrial processes operate at high temperatures, they utilize various extremozymes, for example, glutamate synthetase, Taq and Pfu DNA polymerases, isomerase, amylase, protease, xylanases, pullulanases, glucoamylases, and aminotransferases.

(i) Glutamate synthetase

This enzyme is active at 100 °C and is utilized for the synthesis of glutamine from glutamate and ammonia.

(ii) Taq and Pfu DNA polymerases

Taq DNA polymerase, from *Thermus aquaticus*, is active at 80 °C with pH 8. These DNA polymerases are exploited in polymerase chain reaction (PCR) for amplification of DNA.

(iii) Isomerase

The enzyme, glucose isomerase, converts glucose to fructose and used in food industry as artificial sweetener.

(iv) Amylase

These are used in alcohol, brewing, confectionary, textile, and paper industries. They are extracted from *Pyrococcus furious* and *Pyrococcus woesei*.

(v) Protease

A wide variety of proteases have been isolated, purified, and characterized from *Thermococcus*, *Pyrococcus*, *Staphylothermus*, *Sulfolobus*, and *Desulfurococcus*. A serine protease, pyrolysin, has an optimal temperature of 110 °C with half-life of 4 h at 100 °C. Another serine protease from *Sulfurococcus mucosus* has optimum temperature of 100 °C.

(vi) Pullulanase

Pullulanase II (amylopullulanase) has temperature optima of 105 °C and has been acquired from *Pyrococcus furiosus*, *P. woesei*, *Thermococcus celer*, *T. litoralis*, *Desulfurococcus mucosus*, and *Fervidobacterium pennavorans*. This enzyme has been used in the bioconversion of starch to industrially important products.

(vii) Aminotransferases

Two aromatic aminotransferases have been purified and characterized from *Thermococcus litoralis*. They are found to be active at 100 °C. Aspartate aminotransferase has been isolated from *Sulfolobus solfataricus*. This enzyme transfers amino group from glutamate to oxaloacetate.

5.2 Molecular Chaperones

Molecular chaperones are expressed under stress conditions of high or very low temperatures, and they assist in protein folding without being part of the protein assembly. They have highly conserved sequences, and they are widely spread. They prevent improper aggregation of proteins. Apart from their role in folding newly synthesized proteins, they also have an important role in refolding partially denatured proteins before they are subjected to destruction by protease within the cell. High-temperature molecular chaperones have been isolated from *Sulfolobus solfataricus* and *S. shibatae*.

Due to their ability to survive at high temperatures, thermophilic archaea have been explored and exploited for their use in various bioprocesses, especially for the production of enzymes. The breakthrough research in the field of molecular biology was the identification of Taq polymerase produced by a thermophilic bacterium, *Thermus aquaticus*. Amplification of DNA strands by polymerase chain reaction (PCR) requires a DNA template, DNA polymerase, and primers along with the nucleotide triphosphates and cations. Denaturation of DNA, to separate its two strands, at each step of PCR is performed by subjecting it to a temperature above 90 °C. This step also inactivates DNA polymerase used in the reaction leading to supply of DNA polymerase at each cycle of DNA amplification. This issue has been resolved since the discovery of thermostable Taq polymerase as the optimal temperature range of this enzyme is 75–80 °C with a half-life of 9 min at 97.5 °C. The major drawback of Taq polymerase is the absence of 3′–5′ exonuclease proofreading activity. Another thermostable DNA polymerase, *Pfu* DNA polymerase, from hyperthermophilic archaean, *Pyrococcus furiosus*, overcomes the disadvantages of Taq DNA polymerase in its proofreading activity and removes the wrongly incorporated nucleotides from the newly formed DNA strand.

The second application of thermostable enzymes in molecular biology is in the process of restriction digestion of DNA strands. The conventional Type II restriction enzymes are acquired from mesophilic microorganisms. The problems associated with the mesophilic enzymes are their transport and storage at low temperatures as they are heat-labile. Also, the purification of these enzymes results in low yield due to the loss of the activity of these enzymes at high temperature, thus increasing the cost of these enzymes. These enzymes overcome these disadvantages and provide higher thermostability, freeze-thaw stability, and increased yields during purification. Hence, focus of companies has shifted towards the production of restriction enzymes from thermophilic sources (Sharma et al. 2003).

Several other enzymes produced by thermophilic microorganisms have wide industrial and medical applications. For example, NAD-dependent glutamate dehydrogenase from thermophilic microorganisms has applications as reagents for biosensor development and in bioreactor processes. Amylases and proteases, from thermophilic bacteria and fungi, are applied in starch and peptide fermentations, respectively. Other thermostable enzymes for industrial applications are hydrolases, xylanases, cellulases, and pectinases (Gomes et al. 2016). They also find an application in bioremediation of environment. They can oxidize sulfur and hence are used in various sulfur-reducing, sulfur-utilizing, and sulfur-metabolizing processes. They are also used in hydrogen production, bioleaching, and detoxification of surfaces contaminated with heavy metals. Table 1.3 summarizes the potential applications of different thermophilic archaea.

5.3 Compost: Important Application of Thermophiles

Compost is comprised of degradable organic matter that can be degraded and recycled as fertilizer. They are generally piled up as heap containing sufficient mineral nutrients and aeration for rapid microbial growth. Few examples of compost are the garden compost heap, composts for mushroom production, and composts to process municipal wastes. Figure 1.2 depicts the various steps involved in composting of organic matter.

In the initial phase of a typical composting process, microorganisms grow on the available amino acids and sugars. Mesophilic microorganisms initiate the process. As the process continues, heat is released by their metabolism which increases the temperature of the compost and suppresses the activity of mesophiles. The process is then taken over by some thermophilic fungi and bacteria which raises the temperature of the process to 70–80 °C in a few days. The increased temperature of compost destroys mesophilic organisms and favors the growth of thermophilic organisms. Thermophilic microorganisms grow at peak heating conditions when the temperature is 40–60 °C. Another group of thermophilic fungi *Humicola lanuginosus*, *Humicola insolens*, *Thermoascus aurantiacus*, *Paecilomyces*-like fungus, *Chaetomium thermophilum*, and *Aspergillus fumigatus* start to grow at this stage. Their combined activities result in decomposition of materials present in plant cell wall, namely, cellulose and hemicellulose. This phase of increased temperature reduces the dry weight of compost to half. Finally the temperature of the compost declines which helps mesophilic organisms to recolonize the compost again. This displaces the thermophilic microorganisms; however some heat-tolerant fungal species like *Aspergillus niger* continues to grow as they have growth temperatures from 12 to 55 °C. The fungus with an optimal growth temperature below 50 °C is not considered as a thermophile, but it plays an important role in high-temperature composting.

Table 1.3 List of thermophilic archaea

Type of microorganism	Microorganism	T_{opt} (°C)	Potential products	References
Methanogens—obligate anaerobes	*Methanobacterium thermoautotrophicum*	65	Methane-producing	Zhelkovsky and McReynolds (2012)41
	M. thermalcaliphilum	60		Worakit et al. (1986)
	Methanothermus fervidus	83	D-Glyceraldehyde-3-phosphate dehydrogenase	Stetter et al. (1981) and Fabry and Hensel (1987)
	M. sociabilis	88		Lauerer et al. (1986)
	Methanococcus jannaschii	85	Methane-producing	Jones et al. (1983)
	M. thermolithotrophicus	65		Huber et al. (1982) and Nilsen and Torsvik (1996)118
	M. igneus	88	β-Glutamate	Burggraf et al. (1990a) and Robertson et al. (1990)
	Methanogenium thermophilicum	55		Rivard and Smith (1982)
	Methanopyrus kandleri	98		Kurr et al. (1991)
Thermoacidophiles—obligate and facultative anaerobes	*Thermoplasma acidophilum*	60		Segerer et al. (1988)
	T. volcanium	60	Acid protease	Segerer et al. (1988) and Kocabıyık and Özel (2007)
	Sulfolobus acidocaldarius	75	DNA-dependent RNA polymerase	Brock et al. (1972) and Zillig et al. (1979)
	S. solfataricus	87		Brock et al. (1972)
	Acidianus infernus	90		Segerer et al. (1986)
	Acidianus brierleyi	70		Segerer et al. (1986)

(continued)

Table 1.3 (continued)

Type of microorganism	Microorganism	T_{opt} (°C)	Potential products	References
	Metallosphaera sedula	75	Malonic semialdehyde reductase, succinic semialdehyde reductase, and succinyl-coenzyme A reductase, acetyl-CoA/ propionyl-CoA carboxylase	Huber et al. (1989), Hügler et al. (2003), and Kockelkorn and Fuchs (2009)
	Sulfurococcus mirabilis	73		Golovacheva et al. (1987)
	Desulfurolobus ambivalens	80	Oxygenase reductase	Zillig et al. (1986), Kletzin (1989), and Fuchs et al. (1996)
Thermoneutrophiles—obligate anaerobes	*Thermococcus celer*	88		Zillig et al. (1983b) and Achenbach-Richter et al. (1988)
	T. stetteri	88	Proteinase	Miroshnichenko et al. (1989) and Klingeberg et al. (1995)
	T. litoralis	85	DNA polymerase, amylolytic enzymes, glutamate dehydrogenase	Neuner et al. (1990), Mattila et al. (1991), Brown and Kelly (1993), and Britton et al. (1995)
	Pyrococcus furiosus	100	DNA polymerase, glutamate dehydrogenase, β-glucosidase, hydrogenase, citrate synthase, ferredoxin reductase	Fiala and Stetter (1986), Bryant and Adams (1989), Lundberg et al. (1991), Mukund and Adams (1991), Kengen et al. (1993), Yip et al. (1995), and Russell et al. (1997)

(continued)

Table 1.3 (continued)

Type of microorganism	Microorganism	T_{opt} (°C)	Potential products	References
	P. woesei	102	Inositol phosphate, α-amylase, glyceraldehyde-3-phosphate dehydrogenase, pullulanase, glutamate dehydrogenase, β-galactosidase	Zillig et al. (1987), Zwickl et al. (1990), Koch et al. (1991), Scholz et al. (1992), Ohshima and Nishida (1993), Rüdiger et al. (1995), and Dabrowski et al. (2000)
	Thermoproteus tenax	88		Zillig et al. (1981)
	Thermofilum pendens	88	β-glucosidase, amylolytic enzymes	Zillig et al. (1983a) and Li et al. (2010b, 2013a)
	T. librum	80		Stetter et al. (1986b)
	Desulfurococcus mobilis	85		Zillig et al. (1982)
	D. mucosus	85		Zillig et al. (1982)
	D. amylolyticus	91		Bonch-Osmolovskaya et al. (1988)
	Staphylothermus marinus	92	Amylase, ligase, amylopullulanase	Fiala et al. (1986), Seo et al. (2007), and Li et al. (2010a, 2013b)
	Pyrodictium occultum	105		Stetter et al. (1983)
	P. brockii	105		57
	P. abyssi	97	α-Amylase, endoxylanase	Pley et al. (1991) and Andrade et al. (1996)
	Thermodiscus maritimus	90		58

(continued)

Table 1.3 (continued)

Type of microorganism	Microorganism	T_{opt} (°C)	Potential products	References
	Pyrobaculum islandicum	100	DNA polymerase, NAD-dependent glutamate dehydrogenase, metal reduction	Huber et al. (1987), Kähler and Antranikian (2000), Kashefi and Lovley (2000), and Bhuiya et al. (2005)
	P. organotrophum	100		Huber et al. (1987)
	Archaeoglobus fulgidus	83		Stetter (1988)
	Archaeoglobus profundus	82	Sulfite reductase	Burggraf et al. (1990b) and Larsen et al. (1999)

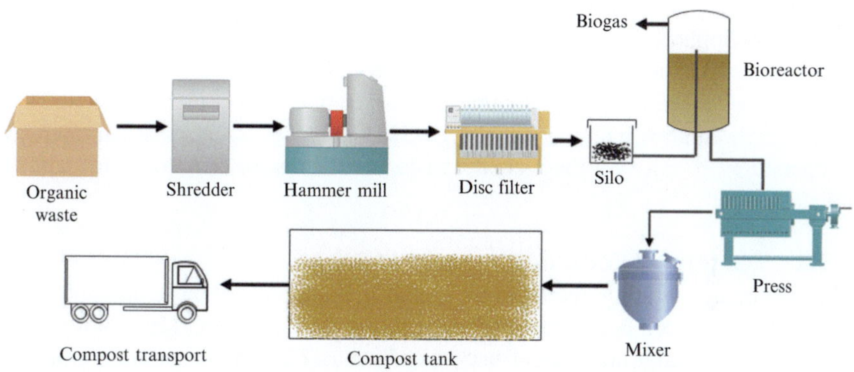

Fig. 1.2 Process flow diagram of composting

5.3.1 Fungi of Composts

Various fungal species can be found in garden composts with temperatures ranging from 20 to 55 °C. They colonize the compost in the initial stages with amino acids and sugars as substrates and are inactivated at later high-temperature stages. *Rhizomucor pusillus* is a gray-colored fungus on potato dextrose agar with aerial hyphae and branched sporangiophores. These spore-bearing stalks or sporangio-phores have sporangia at the tips of their branches. The break in the delicate sporangial walls releases numerous spores. This leaves central bulbous region and the remnants of sporangial wall. *Humicola* (or *Thermomyces*) *lanuginosus* produces single spores at the tips of hyphal branches by a balloon-like swelling process. Upon maturity these spores become brown with ornamented well. This fungus grows at 30–55 °C and can be found in the high-temperature phase of the compost. This organism, however, cannot degrade cellulose and requires cellulose-degrading species for cellulose degradation.

Thermococcus aurantiacus appears as orange-brown-colored colony on malt extract agar which produces small ascocarps (fruiting bodies). Each ascocarp contains eight ascospores with many asci. These ascocarps are closed bodies called as cleistothecia. At maturity the ascospores are released by the breakdown of cleistothecia and ascus walls. This fungus grows at 25–55 °C and degrades cellulose.

Paecilomyces species is a yellow-buff-colored colony on malt extract agar that produces asexual sporing structures on aerial hyphae. The spores of *Paecilomyces* species are formed from flask-shaped cells (phialides) which resemble those of *Penicillium*. These phialides are borne at the tips of short, brushlike branching structures which are less regular than *Penicillium*.

Aspergillus fumigatus is the most common fungus of compost. They are moldy grain and have an asexual sporing stage which consists of spore-bearing hypha called conidiophore. The conidiophore swells and forms a vesicle at the tip. These vesicles bear phialides that produce spores or conidia. The vesicles of *A. fumigatus* are club-shaped with phialides pointing upward from the upper part of vesicle. This organism has a growth temperature of 12–52 °C. As this organism has an optimal temperature of 37 °C, they create major problems in surgeries and cause infections.

5.3.2 Bacteria of Composts

The importance of thermophilic bacteria in degradation of cellulose and polymers in compost has been highlighted by two discoveries. Beffa et al. (1996b) reported the presence of high numbers of bacteria of genus *Thermus* in different types of composts. These bacteria grow on organic substrates at temperatures of 40–80 °C and optimum temperature of 65–75 °C with numbers as high as 107–1010 per gram dry weight of compost. A spore-forming species of *Bacillus* was also reported, but they were unable to survive above 70 °C. It can be conferred that the *Thermus* species, initially isolated from geothermal sites, play an important role during the heating phase of compost. Beffa et al. (1996a) also reported the presence of autotrophic, non-sporing bacteria in composts. These bacteria closely resembled *Hydrogenobacter* strains and can grow at temperatures of 60–80 °C with optimum temperature being 70–75 °C. These microorganisms obtain energy by oxidation of sulfur or hydrogen, utilize carbon dioxide to synthesize organic matter, and were known to be found only in geothermal sites.

6 Biotechnology for Thermophiles: Genetically and Metabolically Engineered Strain Development

The products of thermophiles have gained considerable importance due to the fact that they can withstand high-temperature conditions. This has led to the development of tools for improved and enhanced production of these thermostable enzymes. The reactions utilizing high-temperature bioprocesses offer several advantages when compared to mesophiles. They increase the solubility of substrates which improves the efficiency of conversion reaction. The chances of contamination due to bacteriophages are reduced at high temperatures as well as the greater temperature gradient between ambient air and fermenter decreases the cooling costs (Zeldes et al. 2015).

Genetic engineering or genetic manipulation is the technique to alter genetic makeup of any organism. This can be done by insertion, modification, or deletion of genes in an organism. The organism, thus produced, is known as genetically modified or recombinant organism. There are various techniques for gene manipulation, all of which follow three basic steps. The first step is the isolation of DNA fragments from donor followed by inserting the DNA fragment of donor into a vector. The last step is to grow the recombinant culture in a suitable host (Greenaway 1980). The focus of genetic engineering of thermophiles can be divided into two aspects. In the first approach, the metabolic pathways leading to the formation of undesired products are knocked out. This method eliminates the formation of other fermentation products, and the metabolic pathways of the microorganisms focus on the formation of the compound of interest. The second approach involves insertion of gene producing the desired product into a new organism or enhancing the expression of a gene already present in an organism. Metabolic engineering, on the other hand, is a process to alter the metabolic pathways of organisms leading to an unusual or increased rate of product formation.

Genetically modified microorganisms are utilized for the production of majorly two classes of products: proteins and nonprotein products. Protein products have a wide range of applications in animal husbandry, food processing, industrial catalysts, and therapeutics. The main concern of therapeutics and food processing is the efficiency, authenticity, and safety of the product. Correct posttranslational modifications, in case of therapeutics, and several purification processes to ensure the purity of the products add onto the cost of the products. Nonprotein products are synthesized by metabolic engineering of cells. This can be performed by inserting DNA fragments that code enzymes to generate new or improved pathways for the conversion of precursor substrates to desired metabolites.

A high yield of ethanol was produced from engineered *Thermoanaerobacterium saccharolyticum* which ferments xylan and biomass-derived sugars (Shaw et al. 2008). The organism was knocked out for genes encoding acetate kinase, L-lactate dehydrogenase, and phosphate acetyltransferase which are responsible for

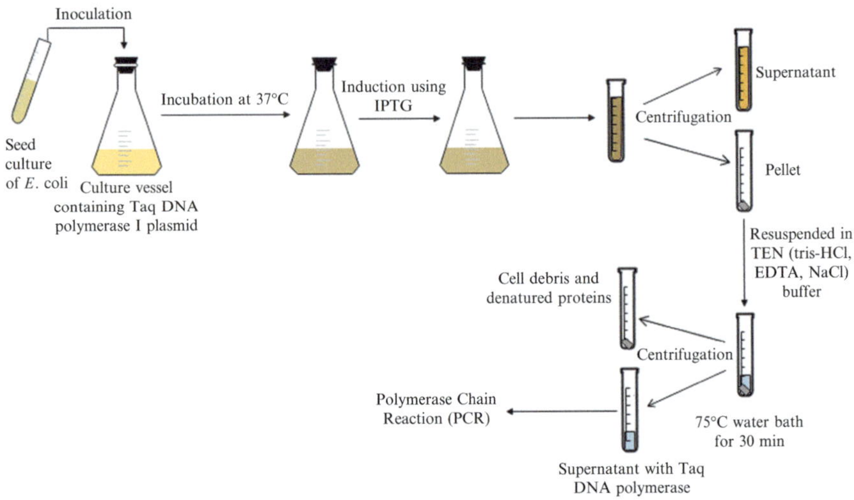

Fig. 1.3 Recombinant production of Taq DNA polymerase in *E. coli*

formation of organic acid. The engineered strain designated as ALK2 utilized pyruvate: ferredoxin oxidoreductase for the formation of ethanol. The ethanol-producing phenotype was found to be stable for more than 150 generations with a yield of 37 g/l. The use of ALK2 for ethanol production at 50 °C ensures simultaneous hydrolysis and fermentation and reduces cellulase as compared to mesophiles. In another study by Argyros et al. (2011), a co-culture of *Clostridium thermocellum* and *T. saccharolyticum* was used for the production of ethanol. The lactate dehydrogenase and phosphotransacetylase genes were deleted by *C. thermocellum hpt* gene and *T. saccharolyticum tdk* gene. The mutant was stable for 2000 h, and the selectivity of ethanol to organic acid with this process was 40:1. The yield of ethanol was 38 g/l from 92 g/l of Avicel fermentation. Both studies demonstrate the potential cost-effective technique of ethanol production from metabolically engineered thermophilic strain. Another thermophilic product which requires recombinant strains for its production is Taq DNA polymerase. Figure 1.3 summarizes the production of recombinant Taq DNA polymerase in *Escherichia coli*. *E. coli* is cultured with Taq DNA polymerase I plasmid which enables the uptake of plasmid into the cell. The induction of cells by IPTG (isopropyl β-D-1-thiogalactopyranoside) leads to the overexpression of the protein of interest (Taq DNA polymerase). The centrifugation, resuspension of pellet, and again centrifugation give the product in the supernatant which can be purified by polymerase chain reaction (PCR) (Protzko and Erickson 2012).

7 Consolidated Bioprocessing

After pretreatment, alcohol production from biomass can be performed by four processes, namely, separate hydrolysis and fermentation (SHF), simultaneous saccharification and fermentation (SSF), simultaneous saccharification and co-fermentation (SSCF), and consolidated bioprocessing (CBP). The process of separate hydrolysis and fermentation (SHF) involves different processes for enzyme production, saccharification, and fermentation. In SSF and SSCF, enzyme production and fermentation are performed separately. In SSF, only hexoses are fermented, whereas SSCF includes fermentation of both pentoses and hexoses. Consolidated bioprocessing is a process in which four biological reactions (enzyme production, saccharification, and fermentation of hexoses and pentoses) are carried out in a single bioreactor. Figure 1.4 depicts a schematic of consolidated bioprocessing. The advantages of consolidated bioprocessing at high temperatures are the requirement of low energy, minimum risk of contamination, improved catalytic ability, and maximum cost-effectiveness (Wiegel 1980).

The microorganisms used in consolidated bioprocessing produce their own enzymes and hence reduce the input cost. The advantages of SSCF are reduced risk of contamination during enzymatic hydrolysis, increased sugar yield, and reduced

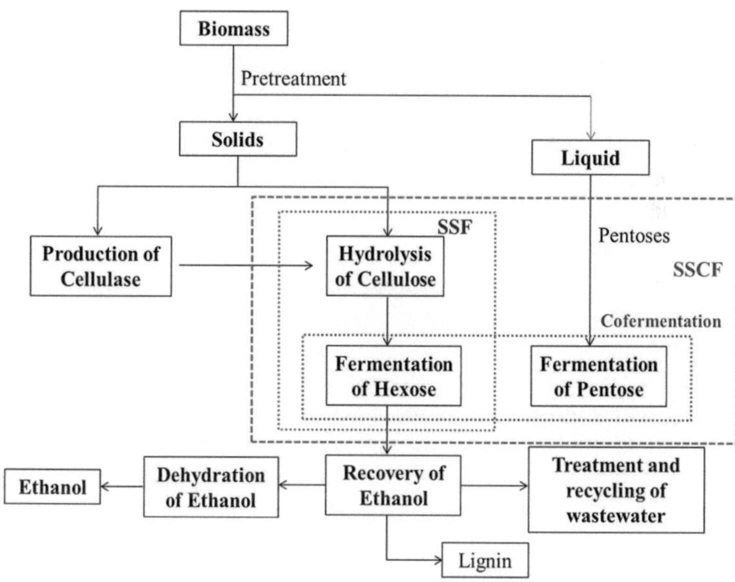

Fig. 1.4 A schematic of consolidated bioprocess

costs of the process. Two approaches are adopted in engineering microorganisms for consolidated bioprocessing. In the first approach, microorganisms degrading lignocellulose are engineered for the efficient production of ethanol. An example of this approach is *Clostridium thermocellum*. In the second approach, ethanol-producing microorganisms are engineered to produce hemicellulases and cellulases, for example, *Saccharomyces cerevisiae*. The microorganisms for consolidated bioprocessing must be resistant to fermentation inhibitors, ethanol, and extreme environments (low pH, high temperature, high osmolarity, and low nutrition capacity) and must be able to utilize both pentoses and hexoses.

Thermoanaerobacterium saccharolyticum is capable of fermenting hemicelluloses and utilizes galactose, glucose, cellobiose, xylose, arabinose, and mannose. The wild-type strain does not ferment cellulose and produce ethanol, lactic acid, and acetic acid as fermentation products. A recombinant strain of *T. saccharolyticum* was developed by Herring et al. (2016) by deleting genes responsible for coding acetate kinase, lactate dehydrogenase, and phosphotransacetylase. The resultant strain produced approximately 60 g/l of ethanol with glucose, xylose, and acetic acid as substrates and SSF as the fermentation condition. This yield was 90% greater than the theoretical yield and shows that the engineered *T. saccharolyticum* can be used in CBP for single-step conversion of purified substrates to ethanol. A cellulose-degrading anaerobic bacterium, *Clostridium* sp. DBT-IOC-C19, was isolated from a Himalayan hot spring by Singh et al. (2017). This bacterium was used for the production of bioethanol by consolidated bioprocessing and converted hemicellulosic and cellulosic substrates to acetate, lactate, and ethanol. Although the ethanol yield produced by this strain was not industrially relevant, the ability of the strain to directly ferment ethanol without any exogenous enzymes can be of potential value. The study also showed the importance of pH and initial substrate concentration in fermentation profile of the strain.

Conventionally, biobutanol is produced by mesophilic strains of bacteria, *Clostridium acetobutylicum* and *C. beijerinckii*. These strains cannot utilize lignocellulose and hence are metabolically engineered to produce butanol from lignocellulose. *Thermoanaerobacterium* sp. M5, a novel thermophilic bacterium, is capable of directly producing butanol from xylan at 55 °C (Jiang et al. 2018). The process involved a single-step consolidated bioprocess utilizing the butanol-ethanol pathway. The expression of alcohol/aldehyde dehydrogenase, β-xylosidase, and xylanase plays an important role in degradation of xylan and synthesis of butanol. Moreover, co-cultivation of *Thermoanaerobacterium* sp. M5 and *C. acetobutylicum* NJ4 increased the butanol titer to 8.34 g/l in CBP as *Thermoanaerobacterium* sp. M5 releases xylose from xylan, and *C. acetobutylicum* NJ4 is a high-butanol producer.

8 Bioprocess Engineering Perspective: Process Development, Modeling, and Optimization

The productivity of microbial processes can be increased by two approaches. In the first approach, inexpensive raw materials are used as alternative substrates for the production of desired compound. This approach also includes the manipulation of biosynthetic routes using mutations or recombinant strains for overproduction. The second approach uses different design, optimization, and modeling techniques for enhancing the performance of bioprocess. These techniques include response surface models, experimental designs, process optimization, and mathematical models for kinetic and process control.

The microbial production of certain biochemical compounds is generally associated with high production costs and results in lower yields. The production costs of these compounds can be increased by employing experimental design methods to reduce the time and effort in planning and executing many experiments. Response surface models help in optimizing the media and process conditions for the production of biochemical and are dependent on data obtained from experimental design. Kinetic models aid in understanding the rates of biochemical reactions, whereas model-based optimization and optimal control studies restrict the optimal operational policies for enhancing the productivity of the product.

Productivity of any microbial product is affected by components of the medium and conditions of the bioprocess. Experimental design helps in designing experiments by varying different experimental conditions simultaneously and studying the influences of these experimental conditions on the formation of product. This reduces the time of experiments and the wastage of chemicals in performing all the experiments. Some of the experimental design procedures are central composite designs, orthogonal designs, and factorial designs. Once the experiments are designed, a relationship is established between the various factors involved and the resulting product which is known as response surface models. Response surface models are used with optimization algorithms to optimize the medium compositions and bioprocess conditions. Artificial neural network and multiple regression models are some of the examples of optimization algorithms with response surface models. The most important studies during the production process are cell growth and conversion of substrate during the bioprocess. These studies are affected by various media compositions and process conditions. Kinetic modeling leads to development of mathematical models that measures the rates of microbial products and predicts the performance of the bioprocess. They also help in the scaling up of the microbial bioprocesses.

In conclusion, other than the generation of recombinant strains for the enhanced production of microbial compounds, modeling and optimization of bioprocesses provide a quicker and cost-effective method for designing an experiment with maximum yield. These techniques reduce the time of operation of a bioprocess and hence are suitable for engineering any bioprocess.

References

Achenbach-Richter L, Gupta R, Zillig W, Woese C (1988) Rooting the archaebacterial tree: the pivotal role of Thermococcus celer in archaebacterial evolution. Syst Appl Microbiol 10(3):231–240

Andrade C, Morana A, De Rosa M Antranikian G (1996) Production and characterization of amylolytic and xylanolytic enzymes from the hyperthermophilic archaeon Pyrodictium abyssi. First International Congress on Extremophiles, Estoril

Aquino A, Jorge J, Terenzi H, Polizeli M (2003) Studies on a thermostable α-amylase from the thermophilic fungus Scytalidium thermophilum. Appl Microbiol Biotechnol 61(4):323–328

Argyros DA, Tripathi SA, Barrett TF et al (2011) High ethanol titers from cellulose by using metabolically engineered thermophilic, anaerobic microbes. Appl Environ Microbiol 77(23):8288–8294

Beeson WT, Iavarone AT, Hausmann CD, Cate JH, Marletta MA (2011) Extracellular aldonolactonase from Myceliophthora thermophila. Appl Environ Microbiol 77(2):650–656

Beffa T, Blanc M, Aragno M (1996a) Obligately and facultatively autotrophic, sulfur-and hydrogen-oxidizing thermophilic bacteria isolated from hot composts. Arch Microbiol 165(1):34–40

Beffa T, Blanc M, Lyon P-F et al (1996b) Isolation of Thermus strains from hot composts (60 to 80 degrees C). Appl Environ Microbiol 62(5):1723–1727

Berka RM, Rey MW, Brown KM, Byun T, Klotz AV (1998) Molecular characterization and expression of a phytase gene from the thermophilic fungus Thermomyces lanuginosus. Appl Environ Microbiol 64(11):4423–4427

Bernhardt G, Lüdemann H-D, Jaenicke R, König H, Stetter KO (1984) Biomolecules are unstable under "black smoker" conditions. Naturwissenschaften 71(11):583–586

Bhuiya MW, Sakuraba H, Ohshima T et al (2005) The first crystal structure of hyperthermostable NAD-dependent glutamate dehydrogenase from Pyrobaculum islandicum. J Mol Biol 345(2):325–337

Bock T, Chen W-H, Ori A et al (2014) An integrated approach for genome annotation of the eukaryotic thermophile Chaetomium thermophilum. Nucleic Acids Res 42(22):13525–13533

Bonch-Osmolovskaya E, Slesarev A, Miroshnichenko M, Svetlichnaya T, Alexeev VA (1988) Characterization of *Desulfurococcus amylolyticus* n. sp.- a new extremely thermophilic archaebacterium isolated from thermal springs of Kamchatka and Kunashir island. Microbiology 57:78–85

Britton KL, Baker PJ, Borges KM et al (1995) Insights into thermal stability from a comparison of the glutamate dehydrogenases from Pyrococcus furiosus and Thermococcus litoralis. FEBS J 229(3):688–695

Brock TD (1967) Life at high temperatures. Science 158(3804):1012–1019

Brock TD, Brock KM, Belly RT, Weiss RL (1972) Sulfolobus: a new genus of sulfur-oxidizing bacteria living at low pH and high temperature. Arch Mikrobiol 84(1):54–68

Brown SH, Kelly RM (1993) Characterization of amylolytic enzymes, having both α-1, 4 and α-1, 6 hydrolytic activity, from the thermophilic archaea Pyrococcus furiosus and Thermococcus litoralis. Appl Environ Microbiol 59(8):2614–2621

Bryant F, Adams M (1989) Characterization of hydrogenase from the hyperthermophilic archaebacterium, Pyrococcus furiosus. J Biol Chem 264(9):5070–5079

Burggraf S, Fricke H, Neuner A et al (1990a) Methanococcus igneus sp. nov., a novel hyperthermophilic methanogen from a shallow submarine hydrothermal system. Syst Appl Microbiol 13(3):263–269

Burggraf S, Jannasch HW, Nicolaus B, Stetter KO (1990b) Archaeoglobus profundus sp. nov., represents a new species within the sulfate-reducing archaebacteria. Syst Appl Microbiol 13(1):24–28

Chapman E, Evans E, Jacobelli M, Logan A (1975) The cellulolytic and amylolytic activity of Papulaspora thermophila. Mycologia 67:608–615

Chen J, Li D-C, Zhang Y-Q, Zhou Q-X (2005) Purification and characterization of a thermostable glucoamylase from Chaetomium thermophilum. J Gen Appl Microbiol 51(3):175–181

Cooney DG, Emerson R (1964) Thermophilic fungi. An account of their biology, activities, and classification. Thermophilic fungi. An account of their biology, activities, and classification

Dabrowski S, Sobiewska G, Maciunska J, Synowiecki J, Kur J (2000) Cloning, expression, and purification of the His (6)-tagged thermostable beta-galactosidase from Pyrococcus woesei in Escherichia coli and some properties of the isolated enzyme. Protein Expr Purif 19(1):107–112

Fabry S, Hensel R (1987) Purification and characterization of d-glyceraldehyde-3-phosphate dehydrogenase from the thermophilic archaebacterium Methanothermus fervidus. FEBS J 165(1):147–155

Fergus C (1971) The temperature relationships and thermal resistance of a new thermophilic Papulaspora from mushroom compost. Mycologia 63(2):426–431

Fernandez-Lafuente R (2010) Lipase from Thermomyces lanuginosus: uses and prospects as an industrial biocatalyst. J Mol Catal B Enzym 62(3–4):197–212

Fiala G, Stetter KO (1986) Pyrococcus furiosus sp. nov. represents a novel genus of marine heterotrophic archaebacteria growing optimally at 100 C. Arch Microbiol 145(1):56–61

Fiala G, Stetter KO, Jannasch HW, Langworthy TA, Madon J (1986) Staphylothermus marinus sp. nov. represents a novel genus of extremely thermophilic submarine heterotrophic archaebacteria growing up to 98 C. Syst Appl Microbiol 8(1–2):106–113

Fuchs T, Huber H, Burggraf S, Stetter KO (1996) 16S rDNA-based phylogeny of the archaeal order Sulfolobales and reclassification of Desulfurolobus ambivalens as Acidianus ambivalens comb. nov. Syst Appl Microbiol 19(1):56–60

Golovacheva R, Zhukova I, Nikultseva T, Ostrovinskii D (1987) Some properties of Sulfurococcus mirabilis, a new thermoacidophilic archaebacterium. Mikrobiologiia 56:281

Gomes E, de Souza AR, Orjuela GL, et al (2016) Applications and benefits of thermophilic microorganisms and their enzymes for industrial biotechnology. In: Gene expression systems in fungi: advancements and applications. Springer, pp 459–492

Gounot A-M (1986) Psychrophilic and psychrotrophic microorganisms. Cell Mol Life Sci 42(11):1192–1197

Greenaway P (1980) Basic steps in genetic engineering. Int J Environ Stud 15(1):11–24

Guerfali M, Gargouri A, Belghith H (2008) Talaromyces thermophilus β-D-xylosidase: purification, characterization and xylobiose synthesis. Appl Biochem Biotechnol 150(3):267–279

Hakulinen N, Kiiskinen L-L, Kruus K et al (2002) Crystal structure of a laccase from Melanocarpus albomyces with an intact trinuclear copper site. Nat Struct Mol Biol 9(8):601

Hakulinen N, Turunen O, Jänis J, Leisola M, Rouvinen J (2003) Three-dimensional structures of thermophilic β-1, 4-xylanases from Chaetomium thermophilum and Nonomuraea flexuosa. FEBS J 270(7):1399–1412

Henriksson G, Akin DE, Slomczynski D, Eriksson K-EL (1999) Production of highly efficient enzymes for flax retting by Rhizomucor pusillus. J Biotechnol 68(2–3):115–123

Herring CD, Kenealy WR, Shaw AJ et al (2016) Strain and bioprocess improvement of a thermophilic anaerobe for the production of ethanol from wood. Biotechnol Biofuels 9(1):125

Huber H, Thomm M, König H, Thies G, Stetter KO (1982) Methanococcus thermolithotrophicus, a novel thermophilic lithotrophic methanogen. Arch Microbiol 132(1):47–50

Huber R, Kristjansson J, Stetter K (1987) Pyrobaculum gen. nov., a new genus of neutrophilic, rod-shaped archaebacteria from continental solfataras growing optimally at 100 C. Arch Microbiol 149(2):95–101

Huber G, Spinnler C, Gambacorta A, Stetter KO (1989) Metallosphaera sedula gen, and sp. nov. represents a new genus of aerobic, metal-mobilizing, thermoacidophilic archaebacteria. Syst Appl Microbiol 12(1):38–47

Huber R, Eder W, Heldwein S et al (1998) Thermocrinis ruber gen. nov., sp. nov., a pink-filament-forming hyperthermophilic bacterium isolated from Yellowstone National Park. Appl Environ Microbiol 64(10):3576–3583

Hügler M, Krieger RS, Jahn M, Fuchs G (2003) Characterization of acetyl-CoA/propionyl-CoA carboxylase in Metallosphaera sedula. FEBS J 270(4):736–744

Jain A (1995) Production of xylanase by thermophilic Melanocarpus albomyces IIS-68. Process Biochem 30(8):705–709

Jatinder K, Chadha B, Saini H (2006) Optimization of culture conditions for production of cellulases and xylanases by Scytalidium thermophilum using response surface methodology. World J Microbiol Biotechnol 22(2):169

Jiang Y, Guo D, Lu J et al (2018) Consolidated bioprocessing of butanol production from xylan by a thermophilic and butanologenic Thermoanaerobacterium sp. M5. Biotechnol Biofuels 11(1):89

Jones W, Leigh J, Mayer F, Woese C, Wolfe R (1983) Methanococcus jannaschii sp. nov., an extremely thermophilic methanogen from a submarine hydrothermal vent. Arch Microbiol 136(4):254–261

Joshi M (1982) A new species of Rhizomucor from India. Sydowia 35:100–103

Kähler M, Antranikian G (2000) Cloning and characterization of a family B DNA polymerase from the hyperthermophilic crenarchaeon Pyrobaculum islandicum. J Bacteriol 182(3):655–663

Kashefi K, Lovley DR (2000) Reduction of Fe (III), Mn (IV), and toxic metals at 100 C by Pyrobaculum islandicum. Appl Environ Microbiol 66(3):1050–1056

Kengen SW, Luesink EJ, Stams AJ, Zehnder AJ (1993) Purification and characterization of an extremely thermostable β-glucosidase from the hyperthermophilic archaeon Pyrococcus furiosus. FEBS J 213(1):305–312

Khademi F, Abachi S, Mortazavi A et al (2013) Optimization of fungal rennet production by local isolate of Rhizomucor nainitalensis. Under solid substrate fermentation system. IOSR J Pharm Biol Sci 5(2):115–121

Kiiskinen L-L, Viikari L, Kruus K (2002) Purification and characterisation of a novel laccase from the ascomycete Melanocarpus albomyces. Appl Microbiol Biotechnol 59(2–3):198–204

Kletzin A (1989) Coupled enzymatic production of sulfite, thiosulfate, and hydrogen sulfide from sulfur: purification and properties of a sulfur oxygenase reductase from the facultatively anaerobic archaebacterium Desulfurolobus ambivalens. J Bacteriol 171(3):1638–1643

Klingeberg M, Galunsky B, Sjoholm C, Kasche V, Antranikian G (1995) Purification and properties of a highly thermostable, sodium dodecyl sulfate-resistant and stereospecific proteinase from the extremely thermophilic archaeon Thermococcus stetteri. Appl Environ Microbiol 61(8):3098–3104

Kocabıyık S, Özel H (2007) An extracellular—pepstatin insensitive acid protease produced by Thermoplasma volcanium. Bioresour Technol 98(1):112–117

Koch R, Spreinat A, Lemke K, Antranikian G (1991) Purification and properties of a hyperthermoactive α-amylase from the archaeobacterium Pyrococcus woesei. Arch Microbiol 155(6):572–578

Kockelkorn D, Fuchs G (2009) Malonic semialdehyde reductase, succinic semialdehyde reductase, and succinyl-coenzyme A reductase from Metallosphaera sedula: enzymes of the autotrophic 3-hydroxypropionate/4-hydroxybutyrate cycle in Sulfolobales. J Bacteriol 191(20):6352–6362

Kunamneni A, Permaul K, Singh S (2005) Amylase production in solid state fermentation by the thermophilic fungus Thermomyces lanuginosus. J Biosci Bioeng 100(2):168–171

Kurr M, Huber R, König H et al (1991) Methanopyrus kandleri, gen. and sp. nov. represents a novel group of hyperthermophilic methanogens, growing at 110 C. Arch Microbiol 156(4):239–247

Larsen Ø, Lien T, Birkeland N-K (1999) Dissimilatory sulfite reductase from Archaeoglobus profundus and Desulfotomaculum thermocisternum: phylogenetic and structural implications from gene sequences. Extremophiles 3(1):63–70

Lauerer G, Kristjansson JK, Langworthy TA, König H, Stetter KO (1986) Methanothermus sociabilis sp. nov., a second species within the Methanothermaceae growing at 97 C. Syst Appl Microbiol 8(1–2):100–105

Leite RSR, Alves-Prado HF, Cabral H et al (2008) Production and characteristics comparison of crude β-glucosidases produced by microorganisms Thermoascus aurantiacus e Aureobasidium pullulans in agricultural wastes. Enzym Microb Technol 43(6):391–395

Li D-C, Lu M, Li Y-L, Lu J (2003) Purification and characterization of an endocellulase from the thermophilic fungus Chaetomium thermophilum CT2. Enzym Microb Technol 33(7):932–937

Li D, Park J-T, Li X et al (2010a) Overexpression and characterization of an extremely thermostable maltogenic amylase, with an optimal temperature of 100 C, from the hyperthermophilic archaeon Staphylothermus marinus. New Biotechnol 27(4):300–307

Li X, Li D, Yin Y, Park K-H (2010b) Characterization of a recombinant amylolytic enzyme of hyperthermophilic archaeon Thermofilum pendens with extremely thermostable maltogenic amylase activity. ppl Microbiol Biotechnol 85(6):1821–1830

Li D, Li X, Dang W et al (2013a) Characterization and application of an acidophilic and thermostable β-glucosidase from Thermofilum pendens. J Biosci Bioeng 115(5):490–496

Li X, Li D, Park K-H (2013b) An extremely thermostable amylopullulanase from Staphylothermus marinus displays both pullulan-and cyclodextrin-degrading activities. Appl Microbiol Biotechnol 97(12):5359–5369

Lloret L, Eibes G, Feijoo G, Moreira M, Lema J (2012) Degradation of estrogens by laccase from Myceliophthora thermophila in fed-batch and enzymatic membrane reactors. J Hazard Mater 213:175–183

Lundberg KS, Shoemaker DD, Adams MW et al (1991) High-fidelity amplification using a thermostable DNA polymerase isolated from Pyrococcus furiosus. Gene 108(1):1–6

Maheshwari R, Kamalam P (1985) Isolation and culture of a thermophilic fungus, Melanocarpus albomyces, and factors influencing the production and activity of xylanase. Microbiology 131(11):3017–3027

Maheshwari R, Bharadwaj G, Bhat MK (2000) Thermophilic fungi: their physiology and enzymes. Microbiol Mol Biol Rev 64(3):461–488

Martins E, Silva D, Da Silva R, Gomes E (2002) Solid state production of thermostable pectinases from thermophilic Thermoascus aurantiacus. Process Biochem 37(9):949–954

Mattila P, Korpela J, Tenkanen T, Pitkämem K (1991) Fidelity of DNA synthesis by the Thermococcus litoralis DNA polymerase—an extremely heat-stable enzyme with proofreading activity. Nucleic Acids Res 19(18):4967–4973

Miettinen-Oinonen A, Londesborough J, Joutsjoki V et al (2004) Three cellulases from Melanocarpus albomyces for textile treatment at neutral pH. Enzym Microb Technol 34(3–4):332–341

Miroshnichenko M, Bonch-Osmolovskaya E, Neuner A et al (1989) Thermococcus stetteri sp. nov., a new extremely thermophilic marine sulfur-metabolizing archaebacterium. Syst Appl Microbiol 12(3):257–262

Moyer CL, Morita RY (2007) Psychrophiles and psychrotrophs. In: Encyclopedia of life sciences. Wiley, Chichester, UK. http://www.els.net

Mukund S, Adams M (1991) The novel tungsten-iron-sulfur protein of the hyperthermophilic archaebacterium, Pyrococcus furiosus, is an aldehyde ferredoxin oxidoreductase. Evidence for its participation in a unique glycolytic pathway. J Biol Chem 266(22):14208–14216

Nakkharat P, Haltrich D (2006) Purification and characterisation of an intracellular enzyme with β-glucosidase and β-galactosidase activity from the thermophilic fungus Talaromyces thermophilus CBS 236.58. J Biotechnol 123(3):304–313

Neuner A, Jannasch HW, Belkin S, Stetter KO (1990) Thermococcus litoralis sp. nov.: a new species of extremely thermophilic marine archaebacteria. Arch Microbiol 153(2):205–207

Nguyen QD, Rezessy-Szabó JM, Claeyssens M, Stals I, Hoschke Á (2002) Purification and characterisation of amylolytic enzymes from thermophilic fungus Thermomyces lanuginosus strain ATCC 34626. Enzym Microb Technol 31(3):345–352

Nilsen RK, Torsvik T (1996) Methanococcus thermolithotrophicus isolated from North Sea oil field reservoir water. Appl Environ Microbiol 62(2):728–731

Ögel Z, Yarangümeli K, Dündar H, Ifrij I (2001) Submerged cultivation of Scytalidium thermophilum on complex lignocellulosic biomass for endoglucanase production. Enzym Microb Technol 28(7–8):689–695

Ohshima T, Nishida N (1993) Purification and properties of extremely thermostable glutamate dehydrogenases from two hyperthermophilic archaebacteria, Pyrococcus woesei and Pyrococcus furiosus. Biosci Biotechnol Biochem 57(6):945–951

Phipps BM, Typke D, Hegerl R et al (1993) Structure of a molecular chaperone from a thermophilic archaebacterium. Nature 361:475–477
Pley U, Schipka J, Gambacorta A et al (1991) Pyrodictium abyssi sp. nov. represents a novel heterotrophic marine archaeal hyperthermophile growing at 110 C. Syst Appl Microbiol 14(3):245–253
Protzko RJ, Erickson FL (2012) A scaled-down and simplified protocol for purifying recombinant Taq DNA polymerase. Bios 83(1):8–11
Puchart V r, Katapodis P, Biely P et al (1999) Production of xylanases, mannanases, and pectinases by the thermophilic fungus Thermomyces lanuginosus. Enzym Microb Technol 24(5–6):355–361
Purkarthofer H, Sinner M, Steiner W (1993) Cellulase-free xylanase from Thermomyces lanuginosus: optimization of production in submerged and solid-state culture. Enzym Microb Technol 15(8):677–682
Rivard CJ, Smith PH (1982) Isolation and characterization of a thermophilic marine methanogenic bacterium, Methanogenium thermophilicum sp. nov. Int J Syst Evol Microbiol 32(4):430–436
Robertson D, Roberts M, Belay N, Stetter K, Boone D (1990) Occurrence of beta-glutamate, a novel osmolyte, in marine methanogenic bacteria. Appl Environ Microbiol 56(5):1504–1508
Rodrigues RC, Fernandez-Lafuente R (2010) Lipase from Rhizomucor miehei as an industrial biocatalyst in chemical process. J Mol Catal B Enzym 64(1–2):1–22
Romdhane IB-B, Fendri A, Gargouri Y, Gargouri A, Belghith H (2010) A novel thermoactive and alkaline lipase from Talaromyces thermophilus fungus for use in laundry detergents. Biochem Eng J 53(1):112–120
Roy SK, Dey SK, Raha SK, Chakrabarty S (1990) Purification and properties of an extracellular endoglucanase from Myceliophthora thermophila D-14 (ATCC 48104). Microbiology 136(10):1967–1971
Rüdiger A, Jorgensen PL, Antranikian G (1995) Isolation and characterization of a heat-stable pullulanase from the hyperthermophilic archaeon Pyrococcus woesei after cloning and expression of its gene in *Escherichia coli*. Appl Environ Microbiol 61(2):567–575
Russell RJ, Ferguson JM, Hough DW, Danson MJ, Taylor GL (1997) The crystal structure of citrate synthase from the hyperthermophilic archaeon Pyrococcus furiosus at 1.9 Å resolution. Biochemistry 36(33):9983–9994
Salar RK, Aneja K (2007) Thermophilic fungi: taxonomy and biogeography. J Agric Technol 3(1):77–107
Schipper MA (1978) On certain species of Mucor with a key to all accepted species. On the genera Rhizomucor and Parasitella. By MAA Schipper, Centraalbureau voor Schimmelcultures
Scholz S, Sonnenbichler J, Schäfer W, Hensel R (1992) Di-myo-inositol-1, 1′-phosphate: a new inositol phosphate isolated from Pyrococcus woesei. FEBS Lett 306(2–3):239–242
Segerer A, Neuner A, Kristjansson JK, Stetter KO (1986) Acidianus infernus gen. nov., sp. nov., and Acidianus brierleyi comb. nov.: facultatively aerobic, extremely acidophilic thermophilic sulfur-metabolizing archaebacteria. Int J Syst Evol Microbiol 36(4):559–564
Segerer A, Langworthy TA, Stetter KO (1988) Thermoplasma acidophilum and Thermoplasma volcanium sp. nov. from solfatara fields. Syst Appl Microbiol 10(2):161–171
Seo MS, Kim YJ, Choi JJ et al (2007) Cloning and expression of a DNA ligase from the hyperthermophilic archaeon Staphylothermus marinus and properties of the enzyme. J Biotechnol 128(3):519–530
Sharma P, D'Souza DR, Bhandari D, Parashar V, Capalash N (2003) Demonstration of the principles of restriction endonuclease cleavage reactions using thermostable Bfl I from Anoxybacillus flavithermus. Biochem Mol Biol Educ 31(6):392–396
Shaw AJ, Podkaminer KK, Desai SG et al (2008) Metabolic engineering of a thermophilic bacterium to produce ethanol at high yield. Proc Natl Acad Sci 105(37):13769–13774
Singh S, Madlala AM, Prior BA (2003) Thermomyces lanuginosus: properties of strains and their hemicellulases. FEMS Microbiol Rev 27(1):3–16

Singh N, Mathur AS, Tuli DK et al (2017) Cellulosic ethanol production via consolidated bioprocessing by a novel thermophilic anaerobic bacterium isolated from a Himalayan hot spring. Biotechnol Biofuels 10(1):73

Souza M d O, Roberto I, Milagres A (1999) Solid-state fermentation for xylanase production by Thermoascus aurantiacus using response surface methodology. Appl Microbiol Biotechnol 52(6):768–772

Srilakshmi J, Madhavi J, Lavanya S, Ammani K (2014) Commercial potential of fungal protease: past, present and future prospects. Int J Pharm Chem Biol Sci 2:218–234

Stetter KO (1988) Archaeoglobus fulgidus gen. nov., sp. nov.: a new taxon of extremely thermophilic archaebacteria. Syst Appl Microbiol 10(2):172–173

Stetter KO (1999) Extremophiles and their adaptation to hot environments. FEBS Lett 452(1–2):22–25

Stetter KO, Thomm M, Winter J et al (1981) Methanothermus fervidus, sp. nov., a novel extremely thermophilic methanogen isolated from an Icelandic hot spring. Zentralbl Bakteriologie Mikrobiol Hyg: I. Abt. Originale C: Allg Angew Ökologische Mikrobiol 2(2):166–178

Stetter KO, König H, Stackebrandt E (1983) Pyrodictium gen. nov., a new genus of submarine disc-shaped sulphur reducing archaebacteria growing optimally at 105 C. Syst Appl Microbiol 4(4):535–551

Stetter KO, Fiala G, Huber R, Huber G, Segerer A (1986a) Life above the boiling point of water? Experientia 42(11–12):1187–1191

Stetter KO, Segerer A, Zillig W et al (1986b) Extremely thermophilic sulfur-metabolizing archaebacteria. Syst Appl Microbiol 7(2–3):393–397

Straatsma G, Olijnsma TW, Gerrits JP et al (1994) Inoculation of Scytalidium thermophilum in button mushroom compost and its effect on yield. Appl Environ Microbiol 60(9):3049–3054

Tan LU, Mayers P, Saddler JN (1987) Purification and characterization of a thermostable xylanase from a thermophilic fungus Thermoascus aurantiacus. Can J Microbiol 33(8):689–692

Tansey MR, Brock TD (1972) The upper temperature limit for eukaryotic organisms. Proc Natl Acad Sci 69(9):2426–2428

Tasca F, Zafar MN, Harreither W et al (2011) A third generation glucose biosensor based on cellobiose dehydrogenase from Corynascus thermophilus and single-walled carbon nanotubes. Analyst 136(10):2033–2036

Ueda S, Udagawa S (1983) Thermoascus-aegyptiacus, a new thermophilic ascomycete. Trans Mycol Soc Jpn 24(2):135–142

Venturi LL, de Lourdes Polizeli M, Terenzi HF, dos Prazeres Melo Furriel R, Jorge JA (2002) Extracellular β-D-glucosidase from Chaetomium thermophilum var. coprophilum: production, purification and some biochemical properties. J Basic Microbiol 42(1):55

Vieille C, Zeikus GJ (2001) Hyperthermophilic enzymes: sources, uses, and molecular mechanisms for thermostability. Microbiol Mol Biol Rev 65(1):1–43

Wiegel J (1980) Formation of ethanol by bacteria. A pledge for the use of extreme thermophilic anaerobic bacteria in industrial ethanol fermentation processes. Experientia 36(12):1434–1446

Worakit S, Boone DR, Mah RA, Abdel-Samie M-E, El-Halwagi M (1986) Methanobacterium alcaliphilum sp. nov., an H2-utilizing methanogen that grows at high pH values. Int J Syst Evol Microbiol 36(3):380–382

Yip K, Stillman T, Britton K et al (1995) The structure of Pyrococcus furiosus glutamate dehydrogenase reveals a key role for ion-pair networks in maintaining enzyme stability at extreme temperatures. Structure 3(11):1147–1158

Zanoelo FF, de Moraes MDLT, Terenzi HF, Jorge JA (2004a) Purification and biochemical properties of a thermostable xylose-tolerant β-D-xylosidase from Scytalidium thermophilum. J Ind Microbiol Biotechnol 31(4):170–176

Zanoelo FF, Polizeli MDLTD, Terenzi HF, Jorge JA (2004b) β-Glucosidase activity from the thermophilic fungus Scytalidium thermophilum is stimulated by glucose and xylose. FEMS Microbiol Lett 240(2):137–143

Zeldes BM, Keller MW, Loder AJ et al (2015) Extremely thermophilic microorganisms as metabolic engineering platforms for production of fuels and industrial chemicals. Front Microbiol 6:1209

Zhelkovsky AM, McReynolds LA (2012) Structure-function analysis of Methanobacterium thermoautotrophicum RNA ligase–engineering a thermostable ATP independent enzyme. BMC Mol Biol 13(1):24

Zillig W, Stetter KO, Janekovic D (1979) DNA-dependent RNA polymerase from the Archaebacterium Sulfolobus acidocaldarius. FEBS J 96(3):597–604

Zillig W, Stetter K, Schäfer W et al (1981) Thermoproteales: a novel type of extremely thermoacidophilic anaerobic archaebacteria isolated from Icelandic solfataras. Zentralbl Bakteriologie Mikrobiol Hyg: I. Abt. Originale C: Allg Angew Ökologische Mikrobiol 2(3):205–227

Zillig W, Stetter K, Prangishvilli D et al (1982) Desulfurococcaceae, the second family of the extremely thermophilic, anaerobic, sulfur-respiring Thermoproteales. Zentralbl Bakteriologie Mikrobiol Hyg: I. Abt. Originale C: Allg Angewandte Ökologische Mikrobiol 3(2):304–317

Zillig W, Gierl A, Schreiber G et al (1983a) The archaebacterium Thermofilum pendens represents, a novel genus of the thermophilic, anaerobic sulfur respiring Thermoproteales. Syst Appl Microbiol 4(1):79–87

Zillig W, Holz I, Janekovic D, Schäfer W, Reiter W (1983b) The archaebacterium Thermococcus celer represents, a novel genus within the thermophilic branch of the archaebacteria. Syst Appl Microbiol 4(1):88–94

Zillig W, Yeats S, Holz I et al (1986) Desulfurolobus ambivalens, gen. nov., sp. nov., an autotrophic archaebacterium facultatively oxidizing or reducing sulfur. Syst Appl Microbiol 8(3):197–203

Zillig W, Holz I, Klenk H-P et al (1987) Pyrococcus woesei, sp. nov., an ultra-thermophilic marine archaebacterium, representing a novel order, Thermococcales. Syst Appl Microbiol 9(1–2):62–70

Zwickl P, Fabry S, Bogedain C, Haas A, Hensel R (1990) Glyceraldehyde-3-phosphate dehydrogenase from the hyperthermophilic archaebacterium Pyrococcus woesei: characterization of the enzyme, cloning and sequencing of the gene, and expression in *Escherichia coli*. J Bacteriol 172(8):4329–4338

Heat-Stable Enzymes from Thermophilic Microorganisms: Biochemical Properties and Commercial Application Potential

2

1 Introduction

Farrell and Campbell classified thermophilic enzymes into three classes (Farrell and Campbell 1969). Class I enzymes are stable at their temperature of synthesis but are inactivated at higher temperatures. Class II enzymes are inactivated at their temperature of synthesis in the absence of substrate. Class III enzymes are those which are heat resistant and stable at temperatures above the temperature of their synthesis (Singleton Jr. and Amelunxen 1973).

Every enzyme is denoted by an Enzyme Commission number depending on the chemical reactions they catalyze. Enzymes are classified into six different groups as mentioned in Table 2.1.

Thermophilic microorganisms tolerate high temperatures of up to 140 °C and still remain active. Because of their higher solvent stability than the mesophilic homologs, they have a wide range of application in biotechnological and biocatalytical industries. The most popular application of these thermostable enzymes is DNA polymerase in polymerase chain reaction (PCR). Thermostability can be useful in purifying thermophilic enzymes as cloning them into mesophilic microorganisms and subjecting these microorganisms to heat will precipitate most of the contaminating mesophilic proteins. This technique has been used in the purification of various enzymes, such as thermostable aldolase from *Sulfolobus solfataricus* (Buchanan et al. 1999).

Thermophilic enzymes withstand high temperature by altering their structure as compared to their mesophilic counterparts keeping the overall structure to be approximately constant. Below 50 °C, that is, at mesophilic temperatures, thermophilic enzymes are rigid compared to their mesophilic homologs. This is due to their compact packing and smaller flexible loop areas. Due to their heat stability and rigidity at mesophilic temperatures, their specific activities are compromised, these being lower at high temperatures as compared to psychrophilic enzymes. On the other hand, at ambient temperatures, they exhibit better stability against denaturants and in organic solvents than their mesophilic homologs. For example, alkaline

J. S. Eswari et al., *Thermophiles for Biotech Industry*,
https://doi.org/10.1007/978-981-32-9919-1_2

Table 2.1 Classification of enzymes based on Enzyme Commission number

EC number	Enzyme classification	Functions	Reaction	Examples
EC 1	Oxidoreductases	Catalyze the transfer of electrons from one molecule to another (oxidation/reduction reactions)	$A^{-}+B \rightarrow A+B^{-}$	Oxidases, dehydrogenases, oxygenases, peroxidases
EC 2	Transferases	Catalyze the transfer of functional groups from one molecule to another	X-group + Y →X + Y-group	Methyltransferases, aminotransferases, kinases, phosphorylases
EC 3	Hydrolases	Catalyze the hydrolysis of hydrogen bonds	A-B + H_2O→ A-OH + B-H	Phosphatases, phosphodiesterases, proteases
EC 4	Lyases	Catalyze the breakdown of various chemical bonds by means other than hydrolysis and oxidation forming a new double bond or ring structure	ATP →cAMP + PP_i	Decarboxylases, aldolases, synthases
EC 5	Isomerases	Catalyze the conversion of a molecule from one isomer to another	A-B→ B-A	Racemases, mutases
EC 6	Ligases	Catalyze the joining of two large molecules by forming a new chemical bond	Ab + C→ A-C + b or Ab + cD →A-D + b + c + d + e + f	Carboxylases, synthetases

α-amylase produced by thermophilic *Bacillus* sp. A3-15 displayed high resistance to sodium dodecyl sulfate (SDS) (Arikan 2008). The global flexibility and rigidity of macromolecules within the cells can be measured using neutron scattering. This technique has shown that macromolecules in thermophiles have lesser flexibility than mesophiles, while psychrophilic macromolecules have the highest flexibility (Tehei et al. 2004; Tehei and Zaccai 2007). Because of this thermophilic enzymes and their diversity are interesting targets for investigation.

2 Oxidoreductases from Thermophilic Hosts

Oxidoreductases, classified as EC (Enzyme Commission) 1, are a group of enzymes that catalyzes the transfer of electrons from one molecule to another. The molecule that transfers the electron is called as reductant (or electron donor), whereas the molecule which receives the electron is called as oxidant (or electron acceptor). For example, the reaction

$$A^{-} + B \rightarrow A + B^{-}$$

will be catalyzed by oxidoreductase. Here, A is the electron donor (reductant) and B is the electron acceptor (oxidant).

There are many flavin-dependent halogenases encoded by different microbial gene clusters. They halogenate many synthetic and natural aromatic compounds which are involved in the biosynthesis of medicinally important halogenated secondary metabolites. The halogenated building blocks provided by the regioselectivity of flavin-dependent halogenases are difficult to get by using standard halogenation chemistries. As a result, flavin-dependent halogenase enzymes are used as biocatalysts for synthesis of pharmaceuticals derived from haloaromatic precursors. However, due to their poor catalytic activity, lack of stability, and poor conversion rates, their application in vitro remains confined. To overcome these issues, a thermophilic tryptophan halogenase was identified from a thermophilic and halotolerant strain of *Streptomyces*. This enzyme was known to significantly improve the catalytic activity and stability compared to the flavin-dependent halogenases known so far. Tryptophan halogenase, in combination with thermostable flavin reductase from *Bacillus*, can halogenate a number of aromatic substrates (Menon et al. 2016).

Kinetic characterization and thermal stability of these enzymes showed higher stability and reaction rate as compared to other halogenases used for biohalogenation in vitro. Crystal structures of these enzymes show an overall structure similar to a mesophilic flavin-dependent halogenase and SttH with similar substrate scope and regioselectivity. Remarkable variations in the amino acid residues on the surface of proteins observed after structural comparisons showed that thermostable tryptophan halogenase has more polar surface residues than SttH. This might reduce aggregation, affect surface hydrogen bonding, and provide higher stability to tryptophan halogenase. These provide an insight for the search of novel, naturally evolved and more stable flavin-dependent halogenases and also in the engineering of these enzymes. The comparison of structure-activity relationships of flavin-dependent halogenases from thermophilic and mesophilic sources can help in improving the properties of these enzymes by targeted mutagenesis from mesophilic microorganisms. The discovery of more stable enzymes can result in effective integration of halogenase biocatalysis with chemocatalytic cross-coupling for scaffold diversification and synthesis of high-value compounds.

Flavoproteins are a group of enzymes that contain a yellow-colored flavin cofactor. They play an important role in many biological processes, namely, degradation of aromatic compounds by hydroxylation, energy synthesis, biosynthesis of ubiquinone, and detoxification of reactive oxygen species. The Old Yellow Enzyme (OYE) family of flavoproteins contains many flavoproteins of unknown functions. Many members of OYE family of flavoproteins catalyze reduction of steroids, but they have not been proven to take part in sterol metabolism.

The enzymes of Old Yellow Enzyme (OYE) family reduce industrially important chemicals. They have been used to study quantum tunneling during enzymatic hydrogen transfer. Only mesophilic homologs within the enzyme family have been studied. Adalbjornsson (2012) characterized Thermophilic Old Yellow Enzyme (TOYE) from a thermophilic bacteria *Thermoanaerobacter pseudethanolicus*. Apart from increased thermostability as compared to mesophilic enzymes, they are stable toward most organic solvents used in industries and hence are industrially favored compared to mesophiles for catalyzing same reactions. The TOYE showed thermal stability up to 70 °C using CD and fluorescence.

Radianingtyas and Wright (2003) reported the presence of alcohol dehydrogenases (ADHs) or similar enzyme in 20 thermophilic archaeal and 17 thermophilic bacterial strains. These thermophiles inhabit hydrothermal vents, hot springs, volcanic marine sediments, solfataric fields, deep-sea chimney walls, manure piles, petroleum reservoirs, and other anthropogenic biotopes. These thermophiles have multiple ADHs which, sometimes, are of different types. Each ADH may have different role or specificity for their survival as the presence of multiple ADHs within an organism suggests the environment in which the organism was exposed.

3 Transferases from Thermophiles

Transferases catalyze the transfer of functional groups from one molecule (donor) to another molecule (acceptor). Transferase-catalyzed reaction can be depicted as:

$$X-\text{group}+Y\rightarrow X+Y-\text{group}$$

In this reaction, X is the donor and Y is the acceptor.

This group of enzymes has an Enzyme Commission number of 2 (EC 2). They have over 450 different enzymes. They transfer single-carbon group, aldehyde or ketone group, acyl groups, glycosyl groups, alkyl groups, nitrogenous groups, phosphorus-containing groups, sulfur-containing groups, selenium-containing groups, and molybdenum- or tungsten-containing groups. They have a variety of roles in cell, for example, peptidyl transferase is used during translation to transfer amino acid and help in peptide chain elongation.

Many of the transferase enzymes are important players in molecular biology applications. Their roles vary from adding nucleotides to the DNA strand (polymerization) to formation of cDNA (reverse transcription). DNA polymerase (EC: 2.7) catalyzes the synthesis of DNA molecules from deoxyribonucleotides. The chemical reaction catalyzed by these enzymes is as follows:

$$\text{Deoxynucleoside triphosphate}+\text{DNA}_{\text{n}}\rightleftarrows\text{diphosphate}+\text{DNA}_{n+1}$$

Thermostable DNA polymerases are major players in DNA amplification and sequencing. Due to the understanding of functionality of DNA polymerase in DNA sequencing and polymerase chain reaction, these methods have been improved recently. Modifying the genes of DNA polymerase can have a potential advantage over its wild counterparts. Such modified enzymes will increase the efficiency and accuracy of DNA amplification process for various applications.

The redesigning of DNA polymerases by introducing single mutations, domain swapping, or tagging has helped in the creation of thermostable DNA polymerases. Tabor and Richardson demonstrated that a single mutation in Taq polymerase from

Thermus aquaticus has opened doors to new sequencing arenas (Tabor and Richardson 1995). These areas have enabled researchers to sequence human genome as well as genome of other organisms. These technologies use genomic sequences and can integrate 400 cycles of amplification with the modified DNA polymerases (Slesarev et al., 2002, Malykh et al., 2004). The incorporation of new thermostable DNA polymerases in sequencing reactions can lead to high processivity and strand displacement ability (Blanco et al., 1989).

Taq polymerase was a major enzyme in PCR amplification of DNA, but this enzyme does not have the ability to amplify damaged DNA. This inability of Taq polymerase restricts its utility in forensic applications. Y-family polymerases, on the other hand, polymerize a wide variety of DNA lesions. McDonald et al. identified five novel Y-family DNA polymerases-like, Dpo4-like, enzymes from *Sulfolobus shibatae*, *Stygiolobus azoricus*, *Acidianus infernus*, *Sulfolobus tengchongensis*, and *Sulfurisphaera ohwakuensis*. These DNA polymerases have the ability to bypass DNA lesions and hence can be used to amplify damaged DNA. They demonstrated that amplifying ultraviolet-irradiated DNA with a blend of Taq and Dpo4-like enzymes resulted in the amplification of DNA with lesions. These thermostable enzymes can be expressed into *E. coli*, and large quantities of this enzyme can be purified from it. The blend of thermostable Y-family polymerase with conventional Taq polymerase has an application in forensic science and in analysis of those samples that cannot be easily amplified by Taq polymerase alone (McDonald et al. 2006).

4 Hydrolytic Enzymes from Thermophilic Microbial Sources

Hydrolase catalyzes the hydrolysis of chemical bonds. They cleave chemical bonds by using water. This cleavage divides a large molecule to smaller molecules. They are represented by the following chemical reaction:

$$A - B + H_2O \rightarrow A - OH + B - H$$

They are classified as EC 3 by the Enzyme Commission. They hydrolyze ester bonds, ether bonds, sugars, peptide bonds, carbon-nitrogen bonds, acid anhydrides, carbon-carbon bonds, halide bonds, phosphorus-nitrogen bonds, sulfur-nitrogen bonds, carbon-phosphorus bonds, sulfur-sulfur bonds, and carbon-sulfur bonds.

Inulinases are β-fructosidases which are used for hydrolysis of inulin, a linear fructose polymer of plant origin. These enzymes were isolated from plants. Thirty-two mesophilic bacterial strains with inulinase activity have been isolated and characterized (Allais et al. 1986). Allais et al. isolated and characterized four thermophilic bacterial strains with inulinase activity. Out of the four thermophilic bacterial strains, namely, LCB36, LCB40, LCB41, and LCB42, identified as belonging to *Bacillus*, *Bacillus* sp. strain LCB41 showed maximum enzyme productivity of 18 U/l/h as compared to 4.0 U/l/h and 1.9 U/l/h for *Kluyveromyces fragilis* and *Candida kefyr*, respectively but in the same range as that of *Aspergillus* species. However, the concentration of enzyme produced by LCB41 was low as compared to the *Aspergillus* sp. These thermostable enzymes have a potential industrial use for inulin or sucrose hydrolysis (Allais et al. 1987).

Lipase is an enzyme that catalyzes the hydrolysis of triglycerides to diglycerides, monoglycerides, glycerol, and fatty acids. Industrial application of lipases includes their use in food, detergent, and pharmaceutical sectors. Thermophilic bacteria can prove to be good candidates for the production of thermophilic lipases. Tayyab et al. isolated a thermophilic microorganism SBS-4S and identified it as belonging to genus *Geobacillus* and similar to *Geobacillus stearothermophilus*, *Geobacillus kaustophilus*, and *Geobacillus uzenensis*. This organism grew at temperatures between 45 and 75 °C and produced amylase, protease, and lipase. The lipase gene when cloned in *Escherichia coli* exhibited optimal activity at 60 °C with stability in the pH range of 6–12. The K_m and V_{max} of this enzyme for the hydrolysis of p-nitrophenyl acetate as calculated from Lineweaver-Burk plot were 3.8 mM and 2273 μmol/min/mg, respectively (Tayyab et al. 2011).

Phytases (myo-inositol hexakisphosphate phosphohydrolase) catalyze the hydrolysis of an indigestible organic phytic acid to usable inorganic phosphorus. They have been commonly detected in fungi. A thermophilic mold *Sporotrichum thermophile* Apinis has been found to produce phytase with sesame oil cake as substrate (Singh and Satyanarayana 2008b). The most significant factors for the production of phytase, namely, glucose, ammonium sulfate, and incubation period, were optimized by Plackett-Burman design. The concentrations of these factors were found to be glucose 3%, ammonium sulfate 0.5%, and incubation period of 120 h. The maximum experimental response for the production of phytase was 362.2 U/g dry moldy residue with an overall improvement of 2.6-fold. In another study, the medium components, such as starch, peptone, Tween 80, and sodium phytate, were optimized for enhanced production of phytase. The concentrations of 0.4%, 0.3%, 1.0%, and 0.3% for starch, peptone, Tween 80, and sodium phytate, respectively, were chosen as the best combination with an increase of 3.73-fold yield of phytase (Singh and Satyanarayana 2008a).

5 Lyases from Thermophiles and Their Applications

Lyase catalyzes the breakdown (elimination) of chemical bonds by means other than substitution of hydrolysis and oxidation. This leads to formation of new double bond or a ring structure. They cleave C-C, C-O, and C-N bonds by elimination reaction, resulting in the elimination of a molecule and the formation of double bond. This enzyme requires one substrate when the reaction proceeds in forward reaction but two substrates in reverse reaction. The representation of a reaction catalyzed by lyase is as follows:

$$\text{ATP} \rightarrow \text{cAMP} + \text{PP}_i$$

They have an Enzyme Commission number of 4 (EC 4). Lyases cleave carbon-carbon bonds, carbon-oxygen bonds, carbon-nitrogen bonds, carbon-sulfur bonds, carbon-halide bonds, and phosphorous-oxygen bonds.

An endospore-forming bacterial strain, later identified as *Bacillus licheniformis* by Singh et al., was isolated from hot extract of sugar beet. The bacterial strain excretes an active exopolygalacturonate lyase with a molecular weight of 38 kDa. The optimum pH of the purified enzyme was 11.0 with an optimum temperature of 69 °C. The enzyme retained its full activity even when kept at 65 °C for 2 h. The action of enzyme yielded 4,5-unsaturated trigalacturonate as product (Singh et al. 1999).

Pectate lyase (EC 4.2.2.2) catalyzes the eliminative cleavage of 1,4-α-D-galacturonan to 4-deoxy-α-D-galact-4-enuronosyl groups. A thermophilic bacterium, *Clostridium stercorarium* F-9, possesses pel9A gene which encodes pectate lyase Pel9A. This enzyme has a molecular weight of 135,171 Da with 1240 amino acids. The enzyme was cloned in *Escherichia coli* and purified. The purified enzyme showed maximum activity at pH of 7.0 with optimal temperature of 65 °C. The enzyme retained 30% of its activity at 50 and 70 °C and was stable after incubation at 70 °C at neutral pH (Hla et al. 2005).

A thermophilic bacterial strain, name proposed as *Defluviitalea phaphyphila* sp., was isolated from a thermophilic alginolytic community. As this specie showed identity with *Defluviitalea saccharophila* LIND6LT2T, the name *Defluviitalea phaphyphila* sp. was proposed by Ji et al. The bacteria possessed six alginate lyases which decompose alginate into unsaturated monosaccharide leading to the production of pyruvate (Ji et al. 2016).

Isocitrate lyase (EC 4.1.3), an enzyme of glyoxylate cycle, catalyzes the breakdown of isocitrate to succinate and glyoxylate. Isocitrate lyase and malate synthase bypass the decarboxylation steps of tricarboxylic acid cycle. Griffiths and Sundaram isolated isocitrate lyase from thermophilic *Bacillus* which showed strong activation (about threefold) by salts. This property is not exhibited by mesophilic microorganisms, *Bacillus megaterium*, *Bacillus licheniformis*, *Escherichia coli*, and *Aspergillus nidulans*. The activation of enzymes by salt is pH dependent that means at pH values above 8.6, KCl inhibits the activity of enzymes. Potassium chloride protects the enzyme from thermal denaturation at 55 °C (Griffiths and Sundaram 1973).

The non-oxidative deamination of L-phenylalanine into (E)-cinnamic acid is catalyzed by phenylalanine ammonia-lyase (PAL; EC 4.3.1.24 and EC 4.3.1.25). In plants, they are required as the starting point of phenylpropanoid pathway, resulting in the biosynthesis of different phenylpropanoids, such as lignins, flavonoids, and coumarins. Because of their role in plant metabolism, they are an important target for herbicides. In addition to this, they also have a therapeutic role in treatment of cancer, in enzyme replacement therapy, and as biocatalyst for the preparation of L- and D-amino acids. Kovacs et al. identified a thermophilic and radiotolerant bacterium, *Rubrobacter xylanophilus*, and sequenced its phenylalanine ammonia-lyase (designated as RxPAL). This enzyme when expressed in *Escherichia coli* TOP 10 exhibited local maximum activity at pH 8.5 and global maximum activity at 11.5. The high pH maximum can be correlated with the presence of disulfide bridges, salt bridges, and excessive negative electrostatic potential on the surface which makes RxPAL a suitable candidate as biocatalyst for formation of L- and D-amino acids (Kovacs et al. 2014).

6 Isomerases from Thermophilic Microorganisms

Isomerases are a group of enzymes that catalyze the conversion of a molecule from one isoform to another. They aid in transferring a specific functional group intramolecularly without forming or breaking any bonds. They have one substrate giving rise to one product with same molecular formula as substrate. Reactions catalyzed by isomerases can be represented as:

$$A - B \rightarrow B - A$$

They have an Enzyme Classification number as EC 5 and catalyze a variety of isomerization reactions, namely, racemization, epimerization, cis-trans isomerization, etc.

Seow et al. purified an alanine racemase (EC 5.1.1.1) from an extremely thermophilic bacteria *Thermus thermophilus* HB8. This enzyme catalyzed the racemization of D- and L-alanine. This enzyme was found to have a molecular weight of 38,000 Da and was active at 75 °C and pH 8. The thermophilic alanine racemase remained active even after its incubation at 80 °C for 30 min. The catalytic activity of this enzyme on other amino acids needs to be investigated. The structure and catalytic mechanism of alanine racemases can be studied by cloning the gene into *Escherichia coli* and overexpressing the enzyme (Seow et al. 2000).

L-Arabinose isomerase (EC 5.3.1.4) catalyzes the reversible isomerization between L-arabinose and L-ribulose. This class of enzyme is called as L-arabinose aldolase-ketose isomerase. These enzymes also catalyze the conversion of D-galactose to D-tagatose. Thermophilic L-arabinose isomerase requires relatively high concentration of metal ions for their activity. Hung et al. cloned araA gene encoding L-arabinose isomerase from *Thermoanaerobacterium saccharolyticum* NTOU1 and expressed it in *Escherichia coli*. The enzyme when purified exhibited maximum activity at 70 °C with pH of 7–7.5. The stability of enzyme was demonstrated in the range of 6.5–9.5, and the enzymatic activity was retained even after 2 h of incubation at 55–65 °C. Also, this isomerase requires lowest metal ion concentration (either 0.1 mM Mn^{2+} or 0.05 mM Co^{2+}). Hence, thermophilic L-arabinose isomerase from *Thermoanaerobacterium saccharolyticum* NTOU1 shows better prospective in industries than the other thermophilic L-arabinose isomerase (Hung et al. 2014).

7 Thermophiles for the Production of Ligases, Molecular Glue of Biotechnology

Ligases catalyze the formation of new chemical bond between two large molecules. This is usually accompanied by hydrolysis of a chemical group on one of the molecules. The general representation of a reaction catalyzed by ligases is:

$$Ab + C \rightarrow A - C + b$$

or

$$Ab + cD \rightarrow A - D + b + c + d + e + f$$

Here, the letters in lowercase signify small, dependent groups.

They are classified as EC 6 by the Enzyme Commission number classification. They join two complementary fragments of nucleic acids and repair single-stranded breaks that arise in DNA during its replication.

Takahashi et al. purified a thermophilic and thermostable DNA ligase from *Thermus thermophilus* HB8. The enzyme had a pH of 6.6 with a molecular weight of 79,000 Da. The DNA ligase had a nick-closing activity up to 85 °C with optimal temperature being 60–85 °C. The enzyme was stable for 1 week at 37 and −20 °C for several months without loss of activity. Also, it can withstand freezing and thawing in 20% glycerol (Takahashi et al. 1984).

8 Recombinant Enzyme Production in Thermophilic Hosts

Trehalose, a nonreducing disaccharide, can be used as sweeteners and seasonings and in preservation of foods and soft drinks. Trehalosyl dextrin-forming enzyme (TDFE) catalyzes intramolecular transglycosyl reaction by converting α-1, 4-glucosidic linkage at the reducing end to α-1, 1-glucosidic linkage. This reaction forms trehalosyl dextrins from dextrin (Fang et al. 2004). Trehalose-forming enzyme (TFE) catalyzes the hydrolysis of α-1, 4 linkages adjacent to the α-1, 1 bond of trehalosyl dextrins to trehalose and low molecular weight dextrins. de Pascale et al. cloned and expressed thermophilic and thermostable TDFE and TFE of *Sulfolobus solfataricus* MT4 in *Escherichia coli*. The yields of these enzymes were 180 U/l and 3630 U/l of medium for TDFE and TFE, respectively. These enzymes can be used for the production of trehalosyl dextrins and trehalose at high temperatures with high yields (de Pascale et al. 2001).

Due to the requirement of new and more efficient enzymes to increase the biochemical reactions, search is focused on native thermophilic enzymes that can be used under industrial conditions. As the maintenance of high temperatures poses a problem in many research and development departments, the simplest solution is to clone and express these enzymes in mesophilic hosts.

Liu et al. isolated a thermophilic bacterial consortium YTY-70 from a hot spring in China. *Caldicellulosiruptor* was the predominant genus in the consortium and was known to produce thermostable β-galactiosidase. This enzyme was cloned and expressed in *Escherichia coli* in fusion with glutathione S-transferase. The recombinant β-galactiosidase showed maximum activity at pH 7 and a temperature of 75 °C. It was highly active for a broad pH range of 5.0 to 9.0 and was stable up to 90 °C. The catalytic activity of this thermostable enzyme was enhanced by Na^+, K^+, Li^+, Mg^{2+}, Ba^{2+}, Ca^{2+}, Fe^{2+}, Zn^{2+}, Mn^{2+}, Cu^{2+}, Al^{3+}, and Fe^{3+}. The enzyme activity was enhanced by Triton X-100, Tween 20, and 2-mercaptoethanol (1 mmol/L) whereas ethylenediaminetetraacetic acid, phenylmethylsulphonyl fluoride, sodium dodecyl sulfate, and 10 mmol/L 2-mercaptoethanol inhibited the activity of the enzyme. The thermophilic β-galactiosidase, when cloned and expressed in mesophilic host, was found to be stable and active with broad ranges of temperature and pH. Hence these recombinant enzymes can be exploited in food industries, especially in dairy industry and for the biosynthesis of galacto-oligosaccharides (Liu et al. 2014).

Roodi et al. cloned and expressed two homologous sequences of amylopullulanases (EC 3.2.1.41) from *Cohnella* sp., *Coh00831* and *Coh01133* in *Escherichia coli* BL21. The optimum temperatures of these enzymes were 70 and 60 °C for *Coh01133* and *Coh00831*, respectively, with optimum pH of 8 and 6. These enzymes exhibited 90% stability in 60 and 50 °C for 90 min, respectively. The metal ions, Mn^{2+}, Ba^{2+}, Ca^{2+}, and $Mg2^+$, increased the enzyme activities for both the enzymes, while metal ions like Fe^{2+}, Li^{2+}, Na2+, Triton X-100, and urea reduced the enzyme activities. The enzymes hydrolyze pullulan to maltotriose and can be used for liquefaction in starch industry (Roodi et al. 2017).

9 Enzyme Kinetics of Thermophilic Microorganisms

Enzyme kinetics is defined as the study of chemical reactions catalyzed by enzymes. The kinetics of single-substrate enzyme-catalyzed reaction was first given by V.C.R. Henri and L. Michaelis and M.L. Menten. Hence, the kinetics of enzyme-catalyzed reactions is termed as Michaelis-Menten kinetics or saturation kinetics. This section briefly summarizes the kinetics of enzyme-catalyzed reaction. The detailed study of kinetics can be found in few reference books (Bailey and Ollis 1976; Shuler and Kargi 2002; Palmer and Bonner 2007).

Due to the fixed number of active sites on enzymes for substrates, these sites become occupied or saturated at high substrate concentrations. Saturation kinetics can be depicted from a simple reaction which involves an enzyme-substrate complex formation (reversible) step and a non-reversible product formation step. This can be represented as:

$$E + S \underset{k_1}{\overset{k_{-1}}{\rightleftarrows}} ES \overset{k_2}{\rightarrow} E + P$$

The Michaelis-Menten equation for the above reaction can be represented as:

$$v = \frac{V_{max}[S]}{K_m + [S]}$$

where v is the initial reaction rate, V_{max} is the maximum rate of enzymatic reaction, [S] is the substrate concentration, and K_m is the Michaelis-Menten constant for the concentration of substrate for which the rate of reaction is half of the maximum velocity of reaction (Michaelis and Menten 1913).

Various modifications of the above Michaelis-Menten equation are used to determine the values of K_m and V_{max}. For example, the equation of double-reciprocal plot or the Lineweaver-Burk plot (Lineweaver and Burk 1934) can be represented as:

$$\frac{1}{v} = \frac{1}{V_{max}} + \left(\frac{K_m}{V_{max}}\right) * \frac{1}{[S]}$$

The slope of the $1/v$ versus $1/[S]$ is linear with slope of K_m/V_{max}.

A plot v versus $v/[S]$ of Eadie-Hofstee plot (Hofstee 1959) results in a slope of $-K_m$ with y-axis intercept as V_m. Eadie-Hofstee plot can be represented as:

$$v = V_{max} - K_m \frac{v}{[S]}$$

The Hanes-Woolf plot represented as equation below helps in determining V_{max} accurately:

$$\frac{[S]}{v} = \frac{K_m}{V_{max}} + \frac{1}{V_{max}}[S]$$

This plot yields a slope of $1/V_{max}$ with K_m/V_m as y-intercept.

Enzyme inhibitors bind to enzymes reversibly or irreversibly and reduce the enzyme activity. Reversible inhibitors can lead to competitive, noncompetitive, and uncompetitive inhibitions. Competitive inhibitors compete with substrate for the active site of enzyme. The scheme of competitive inhibition can be denoted as:

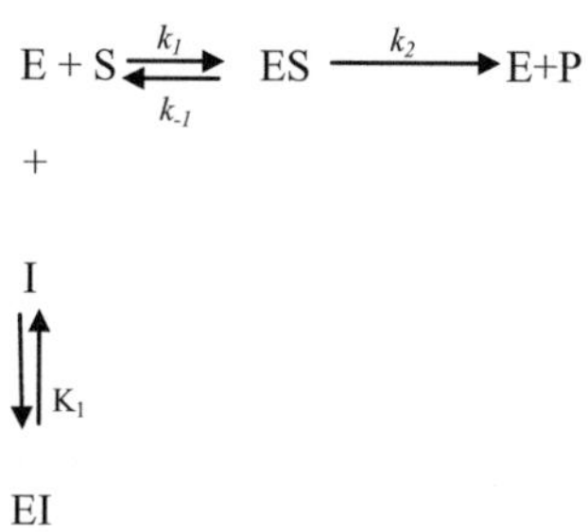

In case of noncompetitive inhibition, the inhibitors do not bind to the active sites of the enzymes but at other sites, thereby reducing the affinity of enzymes to their substrates. These reactions are denoted as:

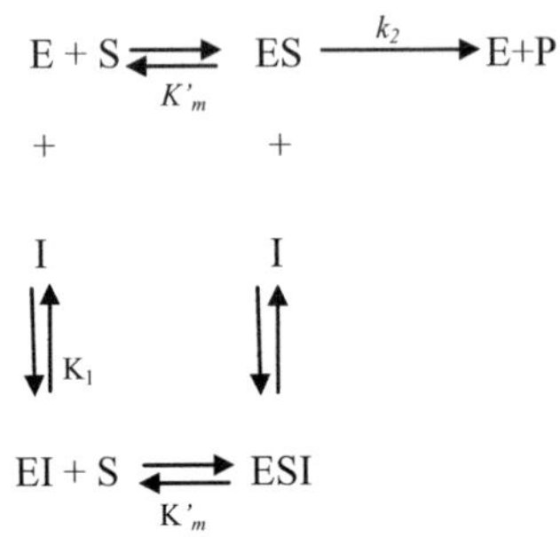

Inhibitors binding only to the ES complex without any affinity for enzyme result in uncompetitive inhibition with the following equation scheme:

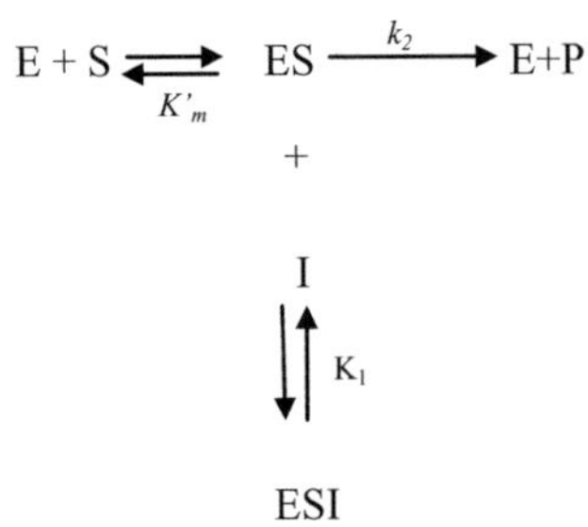

Substrates at high concentrations can also inhibit some enzymatic reactions. These reactions are known as uncompetitive substrate inhibition and can be represented as:

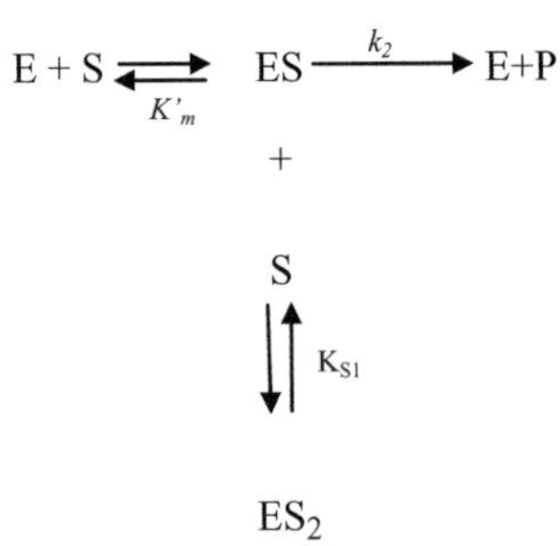

pH plays an important role in enzyme activity. Changes in pH of medium lead to changes in the ionic groups in the active site of enzymes. This results in varied enzymatic activity and hence changes in the reaction rate. Varying pH may also vary the three-dimensional activity of the enzyme, thereby altering the overall enzyme-catalyzed reaction. Temperature is another factor which affects the rate of enzyme-catalyzed reactions. Increasing temperature increases the rate of reaction up to a certain limit. Beyond this optimum temperature, the activity of enzyme decreases due to denaturation of enzyme (Shuler and Kargi 2002; Palmer and Bonner 2007).

The enzyme kinetics in thermophiles depends on several factors. These factors include the stability of substrate at the optimum temperature of enzyme and the state of substrate at such high temperatures. The substrates capable of denaturing at high temperatures will not be available for the enzymes to act on them, and hence the rate of reaction will be reduced. On the other hand, if the substrates are stable at high temperatures, they will be easily available for the enzymes, thereby increasing the overall reaction rate. Another factor which might affect the enzyme kinetics in thermophiles is the state of the substrate. Substrates present in insoluble state will reduce the reaction rate, whereas soluble substrates will increase the rate of reaction. An example of the effect of enzyme kinetics was demonstrated in thermophilic and mesophilic 3-phosphoglycerate kinase by Thomas and Scopes (1998). Their study indicated that the values of k_{cat} increased with increasing temperature but not as rapidly as Arrhenius equation. The maximum k_{cat} values of mesophilic enzyme were seen at temperatures higher than the optimum growth temperature of the mesophilic microorganism, whereas the maximum k_{cat} values of thermophilic enzyme were near the optimal growth temperature. The value of Michaelis-Menten constant was constant throughout the lower temperature but increased at highest k_{cat} values.

10 Immobilization of Thermophilic Enzymes

Enzyme immobilization is defined as the restriction or confinement of enzymes to a matrix or support. The matrix on which enzymes are immobilized helps in exchange of substrate or inhibitor molecules. The carrier matrices are usually inert polymers and inorganic materials. The characteristics of immobilization matrix include stability, inertness, physical strength, regenerability, non-specific adsorption, and ability to reduce product inhibition and increase enzyme activity. Immobilized enzymes offer advantage of increased functionality, enhanced reproducibility, reuse and continuous use of enzymes, minimum reaction time, more stable products, less chances of contamination, improved process, and high enzyme to substrate ratio. On the other hand, immobilization of enzymes has few disadvantages. The isolation, purification, and recovery of active enzymes are costly processes. Some enzymes might become unstable after immobilization and might reduce the catalytic properties.

Enzymes might also become inactivated by heat generated during the process. Enzymes can be immobilized by binding to a carrier, cross-linking, or entrapping inside a polymer matrix. Enzymes can be physically adsorbed on the surface of water-insoluble carriers, such as porous carbon, hydrous metal oxides, polyaromatic resins, glasses, clay, and ion-exchange matrices. The binding of enzyme with carrier can be ionic, hydrogen, covalent, or a combination of these bonds. In covalent bonding, the functional groups of enzymes, namely, amino group, carboxylic group, indole group, imidazole group, and phenol ring, bind to a support matrix. Cross-linking of enzymes forms three-dimensional aggregates of these enzymes. This is favored by covalent binding between the enzyme and the matrix. The enzymes can also be entrapped into polymers like collagen, polyacrylamide, cellulose, and calcium alginate. This technique minimizes the loss of enzyme activity upon immobilization.

Wang et al. demonstrated the immobilization of thermophilic esterase from *Archaeoglobus fulgidus*. The enzyme was covalently coupled with glutaraldehyde after its adsorption on a hydrophobic resin. The immobilized esterase enzyme displayed exceptional operational stability with conversion rate over 90% for a duration of 12 batch reactions and over 80% even after 16 batch reactions (Wang et al. 2014). Ren et al. immobilized thermophilic esterase AFEST from *Archaeoglobus fulgidus*. The immobilization was carried on the epoxy Sepabeads EC-EP as support. The conversion value of the immobilized enzyme was 75% over 15 batch reactions (Ren et al. 2016). Similarly properties of thermophilic proteases, amylases, xylanases, sugar isomerases, glycosidases, lipases, and esterases have been improved upon immobilization (Cowan and Fernandez-Lafuente 2011).

11 Applications of Heat-Stable Enzymes

The thermostable or heat-stable enzymes have immense applications in biotechnological processes. The advantage of these enzymes is that the bioprocesses can be performed at higher temperatures. This enables the reactions to be carried out at higher rates by increasing the diffusion coefficient of substrates and products and increasing the solubility and bioavailability of organic compounds. It also leads to decrease in viscosity of the medium and reduces the chance of contamination by mesophiles (Kumar and Nussinov 2001). Thermostable enzymes can be produced by three ways, namely, thermophilic microorganisms producing thermostable enzymes, mesophilic microorganisms producing thermostable enzymes, and genetically engineered mesophilic microorganisms producing thermostable enzymes. For example, thermostability of enzyme β-glucosidases produced by mesophilic yeast *Aureobasidium pullulans* was compared with thermophilic fungus *T. aurantiacus*. The results showed that the enzyme produced by mesophilic strain was more thermostable than the thermophilic strain (Leite et al. 2007; Ibrahim et al. 2014). Table 2.2 shows some thermostable enzymes along with their applications and microorganisms exploited for their production.

Table 2.2 Applications of heat-stable enzymes, their optimum temperature, applications, and microbial sources

Enzyme	Temperature (°C)	Applications	Microorganism	References
Amylases	50–60	Starch hydrolysis, maltose and dextrin production, bakery industry	*Aeromonas veronii*, *Bacillus* sp., *Stenotrophomonas maltophilia*, *Geobacillus thermoleovorans*, *Thermoanaerobacter finnii*, *Thermobacteroides acetoethylicus*, *Thermoanaerobacter ethanolicus*, *Clostridium thermosaccharolyticum*, *Thermomyces lanuginosus*, *Anoxybacillus flavithermus*, *Thermomonospora vulgaris*	Saito (1973), Allam et al. (1975), Koch et al. (1987), Aiyer (2005), Mishra (2005), Carvalho et al. (2008), Mobini-Dehkordi and Javan (2012), Sen et al. (2014), Rekadwad (2015), and Özdemir et al. (2016)
Cellulases	45–95	Detergent additive and biomass saccharification for ethanol production	*Geobacillus* sp., *Bacillus* sp., *Tepidimonas* sp., *Pyrococcus furiosus*, *Pyrococcus horikoshii*, *Sulfolobus solfataricus*, *Sulfolobus acidocaldarius*, *Sulfolobus shibatae*, *Chaetomium thermophile*, *Sporotrichum thermophile*, *Thermoascus aurantiacus*	Acharya and Chaudhary (2012), Norashirene et al. (2014), Ladeira et al. (2015), and Potprommanee et al. (2017)
Chitinases	70	Bio-control agents against plant pathogens and insects, bioconversion of chitin waste to single-cell protein, oligosaccharides production	*Paenibacillus* sp., *Microbispora* sp., *Staphylococcus* sp., *Thermomyces lanuginosus*	Nawani et al. (2002), Guo et al. (2005), Kuzu et al. (2012), Basu and Chaudhuri (2014), and Chrisnasari et al. (2016)
Laccases	70	Pulp and paper industry and biomass saccharification for ethanol production, decolorization of dye, food industry, baking industry	*Thermobifida fusca*, *Bacillus subtilis*, *Trametes trogii*, *Melanocarpusalbomyces*, *Chaetomium thermophilum*	Chefetz et al. (1998), Minussi et al. (2002), Kiiskinen (2005), Beneyton et al. (2011), Chen et al. (2013), and Yan et al. (2015)
Lichenase (β-1,3; 1,3-Glucanase)	70	Brewing industry, animal feed industry, laundry detergents	*Rhizomucor miehei*, *Sistotrema brinkmannii*, *Malbranchea cinnamomea*, *Aspergillus niger*	Chaari et al. (2012), Tang et al. (2012), Elgharbi et al. (2013), Zhao et al. (2013), and Yang et al. (2014)

(continued)

Table 2.2 (continued)

Enzyme	Temperature (°C)	Applications	Microorganism	References
Lipase	70	Dairy products, detergent additive, biodiesel production, leather industry, and pharmaceutical and fine chemical industries	*Bacillus thermoleovorans*, *Geobacillus* sp., *Anoxybacillus flavithermus*, *Thermosyntropha lipolytica*, *Serratia marcescens*	Lee et al. (1999), Tayyab et al. (2011), Zuridah et al. (2011), Mehta et al. (2012), Burcu Bakir and Metin (2017), and García-Silvera et al. (2018)
β-Mannanase	80	Biomass saccharification for ethanol production and oligosaccharides production	*Neosartorya fischeri*, *Talaromyces leycettanus*, *Bacillus subtilis*, *Thielavia arenaria*, *Dictyoglomus thermophilum*, *Rhizomucor miehei*	Gibbs et al. (1999), Lu et al. (2013), Wang et al. (2015), Yang et al. (2015), Li et al. (2017), and Wang et al. (2018)
Oxidoredutase	70–75	Active alcohols and prochiral ketones in pharmaceutical and fine chemical industries	*Thermoanaerobium brockii*, *Caldicellulosiruptor*	Lamed et al. (1981) and Scott et al. (2015)
Phytase	60	Additive in monogastric feed	*Sporotrichum thermophile*, *Thermomyces lanuginosus*	Berka et al. (1998) and Singh and Satyanarayana (2008c)
Protease	65–85	Baking, brewing, detergents, leather industry	*Coprothermobacter proteolyticus*, *Bacillus* sp.	Nascimento and Martins (2004), Aqel et al. (2012), Panda et al. (2013), and Toplak et al. (2013)
Xylanase	55–105	Pulp and paper industry, textile industry, and biomass saccharification for ethanol production	*B. amyloliquefaciens*, *B. circulans*, *B. pumilus* ASH, *Paecilomyces themophila* J18, *Streptomyces* sp., *Thermoactonomyces thalophilus*, *Thermotoga* sp., *Dictyoglomus* sp., *T. maritima*, *T. neapolitana*, *T. themarum*, *Thermoanaerobacterium saccharolyticum*, *P. furiosus*, *S. solfataricus*, *Thermomyces*, *thermophiles*, *T. lanuginosus*, *Thermoascus aurantiacus*, *Rhizomucor miehei*, *Thermomucor indaticae-seudadicae*	Simpson et al. (1991), Winterhalter and Liebl (1995), Breccia et al. (1998), Dhillon and Khanna (2000), Kohli et al. (2001), Saha (2002), Cannio et al. (2004), Teng et al. (2010), de Lima Damásio et al. (2011), Zhang et al. (2012), and Garg et al. (2013)

References

Acharya S, Chaudhary A (2012) Bioprospecting thermophiles for cellulase production: a review. Braz J Microbiol 43(3):844–856

Adalbjornsson B (2012) Thermophilic old yellow enzyme: structure and kinetic characterisation

Aiyer PV (2005) Amylases and their applications. Afr J Biotechnol 4(13)

Allais J-J, Hoyos-Lopez G, Kammoun S, Baratti JC (1987) Isolation and characterization of thermophilic bacterial strains with inulinase activity. Appl Environ Microbiol 53(5):942–945

Allais J-J, Kammoun S, Blanc P, Girard C, Baratti JC (1986) Isolation and characterization of bacterial strains with inulinase activity. Appl Environ Microbiol 52(5):1086–1090

Allam A, Hussein A, Ragab A (1975) Amylase of the thermophilic actinomycete Thermomonospora vulgaris. Z Allg Mikrobiol 15(6):393–398

Aqel H, Al-Quadan F, Yousef TK (2012) A novel neutral protease from thermophilic Bacillus strain HUTBS62. J BioSci Biotechnol 1(2):117

Arikan B (2008) Highly thermostable, thermophilic, alkaline, SDS and chelator resistant amylase from a thermophilic Bacillus sp. isolate A3-15. Bioresour Technol 99(8):3071–3076

Bailey JE, Ollis DF (1976) Biochemical engineering fundamentals. Chem Eng Educ 19:168

Basu D, Chaudhuri AN (2014) Purification and characterization of chitinase from thermophilic Staphylococcus sp. Int J Environ Sci 4(4):458

Beneyton T, Beyl Y, Guschin DA et al (2011) The thermophilic CotA laccase from Bacillus subtilis: bioelectrocatalytic evaluation of O2 reduction in the direct and mediated electron transfer regime. Electroanalysis 23(8):1781–1789

Berka RM, Rey MW, Brown KM, Byun T, Klotz AV (1998) Molecular characterization and expression of a phytase gene from the thermophilic fungus Thermomyces lanuginosus. Appl Environ Microbiol 64(11):4423–4427

Blanco L, Bernad A, Lázaro JM et al (1989) Highly efficient DNA synthesis by the phage phi 29 DNA polymerase. Symmetrical mode of DNA replication. J Biol Chem 264(15):8935–8940

Breccia JD, Siñeriz F, Baigori MD, Castro GR, Hatti-Kaul R (1998) Purification and characterization of a thermostable xylanase from *Bacillus amyloliquefaciens*. Enzym Microb Technol 22(1):42–49

Buchanan CL, Connaris H, Danson MJ, Reeve CD, Hough DW (1999) An extremely thermostable aldolase from Sulfolobus solfataricus with specificity for non-phosphorylated substrates. Biochem J 343(3):563–570

Burcu Bakir Z, Metin K (2017) Production and characterization of an alkaline lipase from thermophilic Anoxybacillus sp. HBB16. Chem Biochem Eng Q 31(3):303–312

Cannio R, Di Prizito N, Rossi M, Morana A (2004) A xylan-degrading strain of Sulfolobus solfataricus: isolation and characterization of the xylanase activity. Extremophiles 8(2):117–124

Carvalho RVD, Côrrea TLR, Silva JCMD, Mansur LRCDO, Martins MLL (2008) Properties of an amylase from thermophilic Bacillus SP. Braz J Microbiol 39(1):102–107

Chaari F, Bhiri F, Blibech M et al (2012) Potential application of two thermostable lichenases from a newly isolated Bacillus licheniformis UEB CF: purification and characterization. Process Biochem 47(3):509–516

Chefetz B, Chen Y, Hadar Y (1998) Purification and characterization of laccase from Chaetomium thermophilum and its role in humification. Appl Environ Microbiol 64(9):3175–3179

Chen C-Y, Huang Y-C, Wei C-M et al (2013) Properties of the newly isolated extracellular thermo-alkali-stable laccase from thermophilic actinomycetes, Thermobifida fusca and its application in dye intermediates oxidation. AMB Express 3(1):49

Chrisnasari R, Yasaputera S, Christianto P, Santoso VI, Pantjajani T (2016) Production and characterization of chitinases from thermophilic Bacteria isolated from Prataan Hot Spring, East Java. J Math Fundam Sci 48(2):149–163

Cowan DA, Fernandez-Lafuente R (2011) Enhancing the functional properties of thermophilic enzymes by chemical modification and immobilization. Enzym Microb Technol 49(4):326–346

de Lima Damásio AR, Silva TM, dos Reis Almeida FB et al (2011) Heterologous expression of an Aspergillus niveus xylanase GH11 in Aspergillus nidulans and its characterization and application. Process Biochem 46(6):1236–1242
de Pascale D, Sasso MP, Di Lernia I et al (2001) Recombinant thermophilic enzymes for trehalose and trehalosyl dextrins production. J Mol Catal B Enzym 11(4):777–786
Dhillon A, Khanna S (2000) Production of a thermostable alkali-tolerant xylanase from Bacillus circulans AB 16 grown on wheat straw. World J Microbiol Biotechnol 16(4):325–327
Elgharbi F, Hmida-Sayari A, Sahnoun M et al (2013) Purification and biochemical characterization of a novel thermostable lichenase from Aspergillus niger US368. Carbohydr Polym 98(1):967–975
Fang T-Y, Hung X-G, Shih T-Y, Tseng W-C (2004) Characterization of the trehalosyl dextrin-forming enzyme from the thermophilic archaeon Sulfolobus solfataricus ATCC 35092. Extremophiles 8(4):335–343
Farrell J, Campbell L (1969) Thermophilic bacteria and bacteriophages. Adv Microbial Physiol 3:83–109
García-Silvera EE, Martínez-Morales F, Bertrand B et al (2018) Production and application of a thermostable lipase from Serratia Marcescens in detergent formulation and biodiesel production. Biotechnol Appl Biochem 65(2):156–172
Garg G, Dhiman SS, Gautam R et al (2013) Bioscouring of jute fabric by cellulase-free alkalo-thermostable xylanase from Bacillus pumilus ASH. J Mol Catal B Enzym 85:43–48
Gibbs MD, Reeves RA, Sunna A, Bergquist PL (1999) Sequencing and expression of a β-mannanase gene from the extreme thermophile Dictyoglomus thermophilum Rt46B. 1, and characteristics of the recombinant enzyme. Curr Microbiol 39(6):351–357
Griffiths M, Sundaram T (1973) Isocitrate lyase from a thermophilic Bacillus: effect of salts on enzyme activity. J Bacteriol 116(3):1160–1169
Guo R, Li D, Wang R (2005) Purification and properties of a thermostable chitinase from thermophilic fungus Thermomyces lanuginosus. Wei sheng wu xue bao= Acta microbiologica Sinica 45(2):270–274
Hla SS, Kurokawa J, Kimura T, Ohmiya K, Sakka K (2005) A novel thermophilic pectate lyase containing two catalytic modules of Clostridium stercorarium. Biosci Biotechnol Biochem 69(11):2138–2145
Hofstee B (1959) Non-inverted versus inverted plots in enzyme kinetics. Nature 184(4695):1296
Hung X-G, Tseng W-C, Liu S-M, Tzou W-S, Fang T-Y (2014) Characterization of a thermophilic L-arabinose isomerase from Thermoanaerobacterium saccharolyticum NTOU1. Biochem Eng J 83:121–128
Ibrahim D, Weloosamy H, Sheh-Hong L (2014) Potential use of nylon scouring pad cubes attachment method for pectinase production by Aspergillus niger HFD5A-1. Process Biochem 49(4):660–667
Ji S-Q, Wang B, Lu M, Li F-L (2016) Defluviitalea phaphyphila sp. nov., a novel thermophilic bacterium that degrades Brown algae. Appl Environ Microbiol 82(3):868–877
Kiiskinen L-L (2005) Characterization and heterologous production of a novel laccase from Melanocarpus albomyces, VTT Technical Research Centre of Finland
Koch R, Zablowski P, Antranikian G (1987) Highly active and thermostable amylases and pullulanases from various anaerobic thermophiles. Appl Microbiol Biotechnol 27(2):192–198
Kohli U, Nigam P, Singh D, Chaudhary K (2001) Thermostable, alkalophilic and cellulase free xylanase production by Thermoactinomyces thalophilus subgroup C. Enzym Microb Technol 28(7–8):606–610
Kovacs K, Banoczi G, Varga A et al (2014) Expression and properties of the highly alkalophilic phenylalanine ammonia-lyase of thermophilic Rubrobacter xylanophilus. PLoS One 9(1):e85943
Kumar S, Nussinov R (2001) How do thermophilic proteins deal with heat? Cell Mol Life Sci CMLS 58(9):1216–1233
Kuzu SB, Güvenmez HK Denizci AA (2012) Production of a thermostable and alkaline chitinase by Bacillus thuringiensis subsp. kurstaki strain HBK-51. Biotechnol Res Int 2012

Ladeira SA, Cruz E, Delatorre AB, Barbosa JB, Leal Martins ML (2015) Cellulase production by thermophilic Bacillus sp: SMIA-2 and its detergent compatibility. Electron J Biotechnol 18(2):110–115

Lamed R, Keinan E, Zeikus J (1981) Potential applications of an alcohol-aldehyde/ketone oxidoreductase from thermophilic bacteria. Enzym Microb Technol 3(2):144–148

Lee D-W, Koh Y-S, Kim K-J et al (1999) Isolation and characterization of a thermophilic lipase from Bacillus thermoleovorans ID-1. FEMS Microbiol Lett 179(2):393–400

Leite RSR, Gomes E, da Silva R (2007) Characterization and comparison of thermostability of purified β-glucosidases from a mesophilic Aureobasidium pullulans and a thermophilic Thermoascus aurantiacus. Process Biochem 42(7):1101–1106

Li Y-X, Yi P, Yan Q-J et al (2017) Directed evolution of a β-mannanase from Rhizomucor miehei to improve catalytic activity in acidic and thermophilic conditions. Biotechnol Biofuels 10(1):143

Lineweaver H, Burk D (1934) The determination of enzyme dissociation constants. J Am Chem Soc 56(3):658–666

Liu Z, Zhao C, Deng Y, Huang Y, Liu B (2014) Characterization of a thermostable recombinant β-galactosidase from a thermophilic anaerobic bacterial consortium YTY-70. Biotechnol Biotechnol Equip 29(3):547–554

Lu H, Zhang H, Shi P et al (2013) A family 5 β-mannanase from the thermophilic fungus Thielavia arenaria XZ7 with typical thermophilic enzyme features. Appl Microbiol Biotechnol 97(18):8121–8128

Malykh A, Polushin N, Slesarev A, Kozyavkin S (2004) Use of Fimers to eliminate polymerase chain reaction and primer-dimer artifacts and to increase yield in BAC-sequencing reactions. Bacterial Artif Chromosomes: Volume 1 Libr Constr Phys Mapp Sequencing 255:221–229

McDonald JP, Hall A, Gasparutto D et al (2006) Novel thermostable Y-family polymerases: applications for the PCR amplification of damaged or ancient DNAs. Nucleic Acids Res 34(4):1102–1111

Mehta A, Kumar R, Gupta R (2012) Isolation of lipase producing thermophilic bacteria: optimization of production and reaction conditions for lipase from Geobacillus sp. Acta Microbiol Immunol Hung 59(4):435–450

Menon BR, Latham J, Dunstan MS et al (2016) Structure and biocatalytic scope of thermophilic flavin-dependent halogenase and flavin reductase enzymes. Org Biomol Chem 14(39):9354–9361

Michaelis L, Menten M (1913) Die kinetik der invertinwirkung. Biochem Z 49:333–369

Minussi RC, Pastore GM, Durán N (2002) Potential applications of laccase in the food industry. Trends Food Sci Technol 13(6–7):205–216

Mishra RS (2005) Amylases from a thermophilic fungus Thermomyces lanuginosus IISC 91: their purification and properties

Mobini-Dehkordi M, Javan FA (2012) Application of alpha-amylase in biotechnology. J Biol Today's World 1(1):15–20

Nascimento WCAD, Martins MLL (2004) Production and properties of an extracellular protease from thermophilic Bacillus sp. Braz J Microbiol 35(1–2):91–96

Nawani N, Kapadnis B, Das A, Rao A, Mahajan S (2002) Purification and characterization of a thermophilic and acidophilic chitinase from Microbispora sp. V2. J Appl Microbiol 93(6):965–975

Norashirene M, Zakiah Y, Nurdiana S et al (2014) Identification of cellulose-hydrolytic thermophiles isolated from Sg. Klah Hot Spring based on 16S rDNA gene sequence. World Acad Sci Eng Technol Int J Biol Biomol Agric Food Biotechnol Eng 8(9):1041–1044

Özdemir S, Okumus V, Ulutas MS et al (2016) Production and characterization of thermostable α-amylase from thermophilic Anoxybacillus flavithermus sp. nov. SO-19. Starch-Stärke 68(11–12):1244–1253

Palmer T, Bonner PL (2007) Enzymes: biochemistry, biotechnology, clinical chemistry. Woodhead publishing, Oxford

Panda MK, Sahu MK, Tayung K (2013) Isolation and characterization of a thermophilic Bacillus sp. with protease activity isolated from hot spring of Tarabalo, Odisha, India. Iran J Microbiol 5(2):159

Potprommanee L, Wang X-Q, Han Y-J et al (2017) Characterization of a thermophilic cellulase from Geobacillus sp. HTA426, an efficient cellulase-producer on alkali pretreated of lignocellulosic biomass. PLoS One 12(4):e0175004

Radianingtyas H, Wright PC (2003) Alcohol dehydrogenases from thermophilic and hyperthermophilic archaea and bacteria. FEMS Microbiol Rev 27(5):593–616

Rekadwad BN (2015) Characterization of amylase from industrially important thermophilic microorganism: Geobacillus thermoleovorans strain rekadwadsis. Int J Life Sci Biotechnol Pharmacol Res 4(1):26

Ren H, Xing Z, Yang J et al (2016) Construction of an immobilized thermophilic esterase on epoxy support for poly (ε-caprolactone) synthesis. Molecules 21(6):796

Roodi FZ, Aminzadeh S, Farrokhi N, Karkhane A, Haghbeen K (2017) Cohnella amylopullulanases: biochemical characterization of two recombinant thermophilic enzymes. PLoS One 12(4):e0175013

Saha BC (2002) Production, purification and properties of xylanase from a newly isolated Fusarium proliferatum. Process Biochem 37(11):1279–1284

Saito N (1973) A thermophilic extracellular α-amylase from Bacillus licheniformis. Arch Biochem Biophys 155(2):290–298

Scott IM, Rubinstein GM, Lipscomb GL et al (2015) A new class of tungsten-containing oxidoreductase in Caldicellulosiruptor, a genus of plant biomass-degrading thermophilic bacteria. Appl Environ Microbiol 81(20):7339–7347

Sen SK, Raut S, Satpathy S et al (2014) Characterizing novel thermophilic amylase producing bacteria from Taptapani hot spring, Odisha, India. Jundishapur J Microbiol 7(12):e11800

Seow TK, Inagaki K, Nakamura T et al (2000) Purification and some characteristics of a monomeric alanine racemase from an extreme thermophile, Thermus thermophilus. J Biosci Bioeng 90(3):344–346

Shuler ML, Kargi F (2002) Bioprocess engineering: basic concepts, 2nd edn. Prentice Hall, Upper Saddle

Simpson HD, Haufler UR, Daniel RM (1991) An extremely thermostable xylanase from the thermophilic eubacterium Thermotoga. Biochem J 277(2):413–417

Singh B, Satyanarayana T (2008a) Improved phytase production by a thermophilic mould Sporotrichum thermophile in submerged fermentation due to statistical optimization. Bioresour Technol 99(4):824–830

Singh B, Satyanarayana T (2008b) Phytase production by a thermophilic mould Sporotrichum thermophile in solid state fermentation and its potential applications. Bioresour Technol 99(8):2824–2830

Singh B, Satyanarayana T (2008c) Phytase production by Sporotrichum thermophile in a cost-effective cane molasses medium in submerged fermentation and its application in bread. J Appl Microbiol 105(6):1858–1865

Singh SA, Plattner H, Diekmann H (1999) Exopolygalacturonate lyase from a thermophilic Bacillus sp. Enzym Microb Technol 25(3):420–425

Singleton R Jr, Amelunxen RE (1973) Proteins from thermophilic microorganisms. Bacteriol Rev 37(3):320

Slesarev AI, Mezhevaya KV, Makarova KS et al (2002) The complete genome of hyperthermophile Methanopyrus kandleri AV19 and monophyly of archaeal methanogens. Proc Natl Acad Sci 99(7):4644–4649

Tabor S, Richardson CC (1995) A single residue in DNA polymerases of the Escherichia coli DNA polymerase I family is critical for distinguishing between deoxy-and dideoxyribonucleotides. Proc Natl Acad Sci 92(14):6339–6343

Takahashi M, Yamaguchi E, Uchida T (1984) Thermophilic DNA ligase. Purification and properties of the enzyme from Thermus thermophilus HB8. J Biol Chem 259(16):10041–10047

Tang Y, Yang S, Yan Q et al (2012) Purification and characterization of a novel β-1, 3–1, 4-glucanase (lichenase) from thermophilic Rhizomucor miehei with high specific activity and its gene sequence. J Agric Food Chem 60(9):2354–2361

Tayyab M, Rashid N, Akhtar M (2011) Isolation and identification of lipase producing thermophilic Geobacillus sp. SBS-4S: cloning and characterization of the lipase. J Biosci Bioeng 111:272

Tehei M, Franzetti B, Madern D et al (2004) Adaptation to extreme environments: macromolecular dynamics in bacteria compared in vivo by neutron scattering. EMBO Rep 5(1):66–70

Tehei M, Zaccai G (2007) Adaptation to high temperatures through macromolecular dynamics by neutron scattering. FEBS J 274(16):4034–4043

Teng C, Yan Q, Jiang Z, Fan G, Shi B (2010) Production of xylooligosaccharides from the steam explosion liquor of corncobs coupled with enzymatic hydrolysis using a thermostable xylanase. Bioresour Technol 101(19):7679–7682

Thomas MT, Scopes KR (1998) The effects of temperature on the kinetics and stability of mesophilic and thermophilic 3-phosphoglycerate kinases. Biochem J 330(3):1087–1095

Toplak A, Wu B, Fusetti F, Quaedflieg PJ, Janssen DB (2013) Proteolysin, a novel highly thermostable and cosolvent-compatible protease from the thermophilic bacterium Coprothermobacter proteolyticus. Appl Environ Microbiol 79:5625. 01479–01413

Wang C, Luo H, Niu C et al (2015) Biochemical characterization of a thermophilic β-mannanase from Talaromyces leycettanus JCM12802 with high specific activity. Appl Microbiol Biotechnol 99(3):1217–1228

Wang M, Shi H, Wu D et al (2014) Glutaraldehyde cross-linking of immobilized thermophilic esterase on hydrophobic macroporous resin for application in poly (ε-caprolactone) synthesis. Molecules 19(7):9838–9849

Wang X-C, You S-P, Zhang J-X et al (2018) Rational design of a thermophilic β-mannanase from Bacillus subtilis TJ-102 to improve its thermostability. Enzym Microb Technol 97:82

Winterhalter C, Liebl W (1995) Two extremely thermostable xylanases of the hyperthermophilic bacterium Thermotoga maritima MSB8. Appl Environ Microbiol 61(5):1810–1815

Yan J, Chen Y, Niu J, Chen D, Chagan I (2015) Laccase produced by a thermotolerant strain of Trametes trogii LK13. Braz J Microbiol 46(1):59–65

Yang H, Shi P, Lu H et al (2015) A thermophilic β-mannanase from Neosartorya fischeri P1 with broad pH stability and significant hydrolysis ability of various mannan polymers. Food Chem 173:283–289

Yang S, Xiong H, Yan Q, Yang H, Jiang Z (2014) Purification and characterization of a novel alkaline β-1, 3-1, 4-glucanase (lichenase) from thermophilic fungus Malbranchea cinnamomea. J Ind Microbiol Biotechnol 41(10):1487–1495

Zhang F, Hu S-N, Chen J-J et al (2012) Purification and partial characterisation of a thermostable xylanase from salt-tolerant Thermobifida halotolerans YIM 90462T. Process Biochem 47(2):225–228

Zhao Z, Ramachandran P, Choi J-H, Lee J-K, Kim I-W (2013) Purification and characterization of a novel β-1, 3/1, 4-glucanase from Sistotrema brinkmannii HQ717718. J Korean Soc Appl Biol Chem 56(3):263–270

Zuridah H, Norazwin N, Aisyah MS, Fakhruzzaman M, Zeenathul N (2011) Identification of lipase producing thermophilic bacteria from Malaysian hot springs. Afr J Microbiol Res 5(21):3569–3573

Proteins and Amino Acids from Thermophilic Microorganisms: Current Research Trends and Applications

3

1 Introduction

Thermophilic microorganisms synthesize macromolecules with intrinsic thermostability, which does not depend upon any stabilizing factors. Thermophilic organisms survive as a result of the interaction of many mechanisms, namely, rapid growth toward stabilization, more stable membrane, etc. The chief stabilizing factor for survival of thermophile is the heat stability of cellular proteins.

Proteins are the polymers of amino acids and are responsible for physiological function of any biological organisms. Microbial cells consist majorly of proteins, can be grown on inexpensive substrates, and have applications in food and feed additives. Proteins are highly sensitive to temperature and are denatured easily when their optimum temperature is varied. In order for proteins to have the thermal stability, they undergo some modifications in their structures. This chapter discusses the various modifications of proteins for thermal stability as well as the molecular mechanism of the same. Also, the importance of thermostable proteins in industries will be discussed.

2 Thermal Stability of Proteins

Comparing proteins from mesophilic and thermophilic microorganisms helps in understanding the ability of thermophiles to function at high temperatures. This might also provide some signs to define factors that stabilize proteins. Many comparative studies on the thermal stability of proteins have shown that melting temperatures (T_m) of thermophilic proteins are higher than their mesophilic homologs. These comparisons along with detailed thermodynamic analysis will be needed to evaluate the mechanisms employed by thermophilic proteins to remain folded for a wide range of temperatures.

J. S. Eswari et al., *Thermophiles for Biotech Industry*,
https://doi.org/10.1007/978-981-32-9919-1_3

Razvi and Scholtz collected and compiled data from thermophilic and mesophilic protein homologs. They studied the stability curves of many homologous proteins from thermophilic and mesophilic microorganisms which define the variation of conformational stability (DG) with temperature. The conclusion obtained by this compilation was the different methods by which thermophiles thermally stabilize their proteins (Razvi and Scholtz 2006). Thermophilic proteins enhance their thermostability by raising their conformational stability (DG) which increases their T_m. In their study, 77% of thermophilic proteins used higher conformational stability (DG) to increase the melting temperatures (T_m). Another method to attain higher T_m is by lowering the ΔC_p, that is, change in heat capacity associated with protein unfolding. Seventy percent of thermophilic proteins in their study used this technique to lower their T_m. Finally, the last case is where the temperature of maximum stability (T_s; the temperature at which change in entropy between native and denatured state is zero) of thermophilic proteins is higher than the mesophilic homolog. Thirty-one percent of thermophilic proteins considered in their study followed this mechanism.

Many choices are available at the sequence level to increase the DG of a protein as interactions like hydrogen bonds, salt bridges, or hydrophobic interactions add changes in amino acids. Lower ΔC_p can be attained by altering the sequences that provide tighter core packing. Stabilization of proteins by shifting its stability curve to higher temperatures is the least common method of thermostability of proteins as only specific sequences are required to reduce ΔS (change in entropy between the folded and unfolded states). This is because in order to undergo such changes either the entropy of denatured state should be reduced or the entropy of folded state should be enhanced. Majority of the proteins studied by them are not enzymes. This might be because nonenzymatic proteins are more tolerant to enhanced T_s than enzymes.

The thermophilic proteins are stabilized by salt bridges, better core packing, oligomerization, improved secondary structure propensities, tighter loops at the surface, and increased hydrogen bonding. Many studies have compared homologous proteins to determine rules for thermostabilization (Matthews 1993; Vogt and Argos 1997; Vogt et al. 1997; Szilágyi and Závodszky 2000; Petsko 2001) which have proved to be vague. Various computational tools have been used to design proteins with increased thermostability (Dahiyat et al. 1997; Malakauskas and Mayo 1998; Korkegian et al. 2005). Directed evolution can be used to design proteins without any prior knowledge of the stabilizing/destabilizing factors (Sieber et al. 1998; Arnold et al. 2001; Wintrode and Arnold 2001; Eijsink et al. 2005).This technique provides a high rate of success as proteins depend on combinations of stabilizing factors rather than a single molecular or thermodynamic effect to function at higher temperatures. This depicts the adaptability of proteins and to maintain their structure and functions at high temperatures.

Melchionna et al. studied the relation of thermostability of proteins with water. They showed that water destabilizes the protein by forming hydrogen bonds but the water molecules present in the proximity of the molecule act as a bioprotectant and protect the molecule against chemical, mechanical, and thermal stress. Hence, thermophilic proteins employ the strategy of flooding the protein core with water to allow small number of water molecules to penetrate the protein-water interface. This layer acts as a barrier and does not allow penetration of other solvent inside the protein core. The interaction between protein and water is also one of the reasons for thermostability of thermophilic proteins (Melchionna et al. 2006).

Sawle and Ghosh studied the thermostability of 116 proteins by analyzing the entropy and enthalpy at convergence temperature. It was found that enthalpy gain upon folding is lower in thermophiles than mesophiles, and entropy loss upon folding is higher in mesophiles than thermophiles. This suggests the role of entropic stabilization for high melting temperature. It was observed that specific enthalpy and entropy of thermophilic proteins exhibited fewer changes upon folding. Also, the free energy of protein folding at a temperature of maximal stability is more favorable in thermophilic microorganisms than mesophilic microorganisms. These results suggest that alteration of protein thermodynamics helps thermophiles to achieve high melting temperatures and to adapt to high temperatures as compared to mesophiles (Sawle and Ghosh 2011).

3 Factors Affecting Thermal Stability of Proteins

Many theoretical and experimental investigations had led to the knowledge of thermostability of thermophilic proteins. Thermostability of proteins can be obtained by many small structural modifications rather than amino acid substitutions. Exchange of amino acids is observed in thermophilic proteins as compared to its mesophilic homologs. Theoretical studies include comparative investigation of thermophilic and mesophilic proteins to expose their sequences and structural differences that lead to the difference in thermostability properties. Sadeghi et al. compared thermophilic proteins with their mesophilic homologs. This strategy provides the advantages of eliminating noise or background from signals along with recognizing the factors that lead to the stability of thermostability of proteins. In both the thermophilic and mesophilic proteins, polar and nonpolar residues contributed equally to surface area and compactness of proteins, whereas thermophilic proteins show an increase in salt bridges and main-chain hydrogen bonds than their mesophilic homologs. Moreover, thermophilic proteins have a frequent occurrence of hydrophobic residues and less of polar residues. These indicate that thermostable proteins have undergone several evolutionary strategies to withstand high temperatures (Sadeghi et al. 2006).

The reasons for increased stability in thermophilic proteins are greater hydrophobicity of core, tighter packing, deletion or shortening of loops, smaller voids or cavities, increased occurrence of proline residues, substitution of amino acids within and outside the secondary structure, increased helical content, increased polar surface area, increased hydrogen bonding and salt bridges, increased surface area buried upon oligomerization, higher isoelectric point, and decreased occurrence of

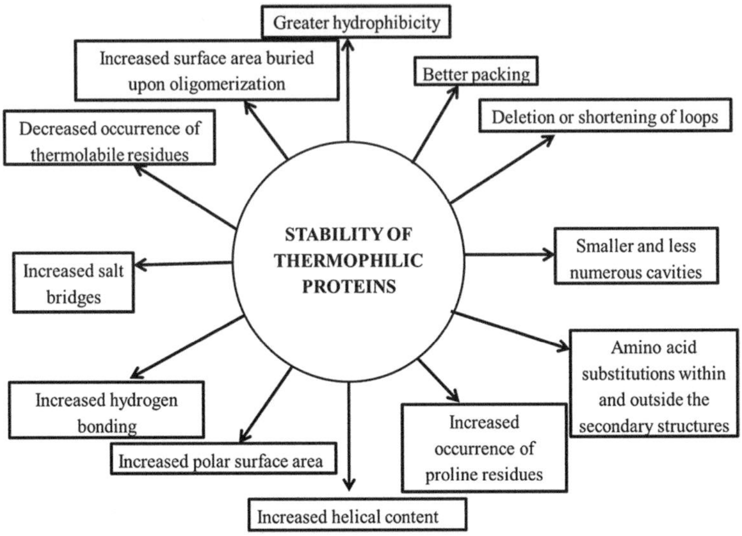

Fig. 3.1 Factors affecting stability of thermophilic proteins

thermolabile residues (Fig. 3.1) (Taylor and Vaisman 2010). Few of the amino acids, namely, asparagine, glutamine, methionine, and cysteine, are thermolabile, that is, they are not stable at high temperatures. These amino acids undergo deamidation (asparagine and glutamine) or oxidation (methionine and cysteine) and hence are less common in thermophilic proteins (Kumar et al. 2000). The amino acids, glycine, lysine, tyrosine, and isoleucine, are the preferred amino acids in thermophilic microorganisms, whereas glycine, histidine, alanine, and cysteine are favored in mesophiles (Cambillau and Claverie 2000; Farias and Bonato 2003). Haney et al. showed that amino acids, arginine, glutamine, isoleucine, lysine, and tyrosine, occur at a higher frequency and asparagine, glutamine, glycine, methionine, serine, and threonine occur at lower frequencies (Haney et al. 1999).

Kumar et al. observed the similarities in compactness, polar and nonpolar residues, hydrophobicities, and main-chain and side-chain hydrogen bonds of both thermophilic and mesophilic proteins, whereas insertions/deletions and proline substitutions are not consistent between the two groups of proteins. Increased frequencies of salt bridges and side-chain hydrogen bonds were observed in thermophilic proteins. Thermophilic proteins exhibited higher occurrence of arginine and tyrosine residues with a lower occurrence of cysteine and serine residues. Also, they had a higher percentage of residues in α-helical conformation avoiding proline residues in their helices compared to mesophiles. These strategies help thermophilic proteins to tolerate high temperatures (Kumar et al. 2000).

4 Phenomena of Protein Folding System in Thermophiles

Protein folding is a process by which a polypeptide folds into a stable and functional three-dimensional structure. By this process, the protein structure adopts a functional conformation or shape. Many of the proteins are stabilized and functionalized by disulfide bonds. Various thiol-disulfide oxidoreductases are known which aid in the formation of disulfide bonds in proteins in all living cells. These thiol-disulfide oxidoreductases belong to thioredoxin family. Polypeptides are in random coil conformation when they come out of ribosome and in a linear form in cytosol. Folding of these proteins requires their recognition by chaperones which prevent them from coiling until they enter the folding pathway. This function is performed by a chaperone named trigger factor (TF) in bacteria and nascent polypeptide-associated complex (NAC) in archaea. These proteins prevent interaction of polypeptide chains with other cellular components.

Organisms can acquire thermotolerance when they are exposed to heat shock. Heat shock proteins (HSPs) are a group of effector proteins which protect proteins from heat stress. These proteins perform the functions of molecular chaperones, catalyze the refolding of denatured proteins, assist folding of new proteins, or prevent protein aggregation. Small HSP or sHSP proteins have a molecular weight of 15–40 kDa. Holdases, a class of sHSP from extremophiles, prevent denatured proteins from aggregation and refold denatured proteins (Luo and Robb 2011). The number of sHSP genes is variable in archaeal species ranging from one to three homologs. Hyperthermophiles, growing at 100 °C, have one *shsp* gene except *Pyrobaculum aerophilum* which has two homologs (Laksanalamai et al., 2001), whereas *Thermoplasma acidophilum* have three *shsp* homologs (Ruepp et al. 2001).

The misfolded proteins in thermophilic archaea and bacteria are disposed by a large hollow complex termed as 20S proteasome. They are similar to eukaryotic 20S proteasome with few proteins which makes them a simple system. Eukaryotes have 23 unique proteasome subunits, whereas *T. acidophilum* consist of only two proteasome subunits (Pühler et al. 1992; Zwickl et al. 1992). Similar to eukaryotic proteasomes, the archaeal proteasome also consists of a stack of four heptameric rings forming a barrel-like structure where protein degradation occurs (Rechsteiner et al. 1993).

5 Molecular Mechanisms of Action of Thermostable Proteins

At high temperatures at which thermophiles and hyperthermophiles grow, proteins undergo unfolding. Due to this unfolding, their hydrophobic core is exposed, leading to protein aggregation (Tomazic and Klibanov 1988). Thermophilic proteins have adapted to retain their structure and function at extreme temperatures. These adaptations include more number of disulfide bonds, large hydrophobic residues, and ionic interactions.

5.1 Disulfide Bonds

Disulfide bonds between cysteine residues are an important determinant in the overall structure of protein. These bridges either activate or inactivate enzymes. These bonds play an important role in thermostable enzyme by increasing the stability of these proteins and preventing change in quaternary structures. The enzyme 5′-deoxy-5′ methylthioadenosine phosphorylase II is thermostabilized by disulfide bridges (Cacciapuoti et al. 1994, 2012). This enzyme was used to study intrasubunit disulfide bonds in thermophilic proteins. It was found using circular dichroism that under reducing conditions, this hexameric protein disassociated into a reversible monomeric states, whereas denaturation chemically and thermally resulted in irreversible change in structure. Mutations within CXC were shown to decrease thermal stability by interfering in disulfide bridging (Cacciapuoti et al. 2012). The CXC motif forms 11-member disulfide ring (Woycechowsky and Raines 2003). This is similar to disulfide isomerases which help in the maintenance of proper disulfide bridges within proteins (Wilkinson and Gilbert 2004).

Disulfide bonds are also required for oligomerization. Boutz et al. (2007) demonstrated that disulfide bonds in citrate synthetase from *Pyrobaculum aerophilum* created cyclized protein chains by linking two monomeric subunits of the homodimer. The structure provides stability to the dimer by preventing the individual subunits from separating. The examples mentioned above depict the role disulfide bonds play in increasing the thermostability of proteins by either increasing rigidity or interlocking adjacent chains in monomeric subunits (Reed et al. 2013).

5.2 Salt-Bridging

Salt bridges destabilize mesophilic proteins and are not favorable when compared to hydrophobic interactions (Hendsch and Tidor 1994). On the other hand, salt bridges are a predominant feature of thermophilic enzymes (Karshikoff and Ladenstein 2001). At higher temperatures, entropic cost associated with ion pairing in salt bridges and desolvation penalty are easily overcome (Chan et al. 2011). Negation of these considerations leads to salt bridges becoming the structurally stabilizing element. This increases thermal capacity of proteins by favorable charge-charge interactions.

Circular dichroism studies of thermophilic ribosomal protein, L30e from *Thermophilus celer*, showed some change in thermal capacity of protein without causing any structural change. Mutations in residues involved in salt-bridging to hydrophobic residues lead to increase in heat capacity change of unfolding.

Lowering this heat capacity change can increase thermostability of proteins which prefers natively folded state over unfolded state. This indicates that salt bridges improve thermal stability of proteins (Chan et al. 2011).

5.3 Large Hydrophobic Core and Oligomerization

Thermophilic proteins have shown to have some deviations from standard quaternary organization. This deviation seems to increase the rigidity of subunits of hydrophobic core, thereby promoting tighter packing and reducing the exposure of hydrophobic residues to solvent (Vieille and Zeikus 2001). This phenomenon can be explained by the characterization of two acetyl-CoA synthetases from *Ignicoccus hospitalis* (Mayer et al. 2012) and *Pyrobaculum aerophilum* (Bräsen et al. 2005). These acetyl-CoA synthetases have different oligomeric state than their mesophilic counterparts. The thermophilic enzymes form octomers, whereas the mesophilic counterparts form monomers or homodimers. The acetyl-CoA synthetases from *Archaeoglobus fulgidus*, a lower growth temperature hyperthermophile, form a trimer (Ingram-Smith and Smith 2006).

With increasing temperature the effect of hydrophobicity becomes important in protein stability and folding. Tighter packing in phosphotriesterase from *Sulfolobus solfataricus* has been seen due to favorable hydrophobic interactions at dimer interface (Del Vecchio et al. 2009). This technique can also be taken up by hyperthermophilic acetyl-CoA synthetases to decrease surface area to volume ratio with respect to individual subunits and solvent-exposed hydrophobic regions. This leads to tighter packing of hydrophobic core representing a general feature of thermostable proteins. On the other hand, increasing the oligomeric state in acetyl-CoA synthetases from *Pyrococcus furiosus* exhibits lack of oligomerization when compared to mesophiles (Park et al. 2013). This depicted as the first example of a monomeric cyclodextrin-hydrolyzing enzyme. Bacterial enzymes need to be dimerized before their activity (Park et al. 2000). The N′-terminus domain of *Pyrococcus furiosus* amylase makes it active as a monomer. In bacteria, this domain consists of a loop which spreads over the active site acting as lid and stabilizes some substrates like maltose, whereas this is not the case in *Pyrococcus furiosus* amylase.

Even though it is thought that structural flexibility of proteins is not favored in thermostable enzymes, no mechanism has been shown by which this rigidity is achieved (Vihinen 1987). Higher oligomeric state is favored in thermophilic acetyl-CoA synthetase, whereas in *Pyrococcus furiosus* amylase, all components are organized as a single subunit which creates structural rigidity and tightens the packing of hydrophobic core. Both hypotheses show the advantages of changes in quaternary structure.

5.4 Surface Charges

Increased surface charges on the surface of proteins are abundantly present in thermostable proteins (Fukuchi and Nishikawa 2001). The stability of proteins can be increased by replacing polar uncharged residues with polar charged residues. The amino acids, such as asparagine and glutamine, undergo deamination at higher temperatures which reduce stability of proteins. Replacement of these residues increases charged interactions, thereby protecting them against thermal degradation (Lee et al. 2005).

Lee et al. created multiple single-point mutations on ribosomal protein, L30e from *Thermophilus celer*, at places other than those involved in salt bridges. The thermal capacity of L30e can be increased by mutations in charged residues and can be decreased by replacing surface charges with alanine. Long charge-charge interactions resulted in stable L30e as removing these interactions resulted in greater chance of thermal and chemical denaturation (Lee et al. 2005). While presence of higher surface charged residues stabilizes the proteins by preventing their aggregation at higher temperatures, it also destabilizes the protein structure (Mamat et al. 2002). This suggests that the extremely charged surfaces require compensation for its structural adaptation.

6 Significance of Thermophilic Proteins at Industrial Level

The optimal activities of thermophilic proteins at high temperatures make them suitable for biotechnological and industrial applications that require the process to be performed at high temperatures. At such high temperatures, the kinetics and thermodynamics of catalyzed reactions are favorable (Unsworth et al. 2007) which permit reaction to be efficient and result in higher product yield. The other benefits of thermostable proteins and enzymes are the reduced chance of bacterial contamination, especially in food and drug applications and reduced operating costs from enzyme replacement due to thermal denaturation. The applications of thermophilic enzymes started with Taq DNA polymerase from *Thermus aquaticus*. This enzyme reduced the cost and allowed the process of polymerase chain reaction (PCR) which led to advanced research in molecular biology (de Champdoré et al. 2007; Unsworth et al. 2007). Currently, other than *T. aquaticus*, DNA polymerases (pfuTurbo, DeepVent$_R$, Therminator) from different thermophilic archaeal species are used. DNA ligases active at 45–80 °C have also proven to be beneficial in molecular biology to ligate oligonucleotides in DNA. These are produced by *P. furiosus* (Pfu DNA ligase) and *Thermus scodoductus* (Tcs DNA ligase). Apart from their applications in molecular biology, they are also used in various industrial applications. Hydrolysis of starch requires the processes of liquefaction and saccharification to be performed at high temperatures. This process requires use of α-amylase, β-amylase, pullulanase, glucoamylase, and isoamylase. The enzymes are obtained from mesophilic microbial sources and hence are stable below 60 °C. Saccharification at higher temperatures involves low enzyme purification and pumping costs, increased substrate concentration and reaction rates, lower risk of bacterial contamination, and lower half-life of catalyst. In order to achieve these, thermophilic enzymes are employed in starch processing industry. Thermophilic α-amylase is produced by *P. furiosus*, *P. woesei*, *Desulfurococus mucosus*, *Pyrodictum abyssi*, *Staphylothermus marinus* with optimum activity at 100 °C, *Dictyoglomus thermophilum* at 90 °C, *Thermotoga maritima* at 85–90 °C, and *Thermococcus profundus* at 80 °C. β-amylases are produced by *T. maritima* and *Thermoanaerobacterium thermosulfurigenes* with optimum activities at 95 °C and 75 °C, respectively (Vieille and Zeikus 2001).

Park et al. demonstrated that mutation in a thermostable amylase from archaeal *Pyrococcus furiosus* led to increased production of maltohepatose from β-cyclodextrin. The compound maltoheptose and maltooligosaccharides have a high potential in food, pharmaceutical, and cosmetic industries (Park et al. 2013). Other applications of thermophilic enzymes are as optical nanosensors. In this case, due to the lack of catalytic activity of these enzymes at ambient temperatures, they bind to substrate but do not convert it to product. The enzyme-substrate complex formed can be identified by evaluating difference in enzyme fluorescence which allows for easy quantification of the amount of substrate in the sample. These techniques have important prospective in biotechnology, drug discovery, and medical testing (de Champdoré et al. 2007).

7 Bioprocess Considerations for Amino Acids and Proteins from Thermophilic Microorganisms

The production of amino acids using bioprocess technology was developed in the end of 1950s. Initially, this technology was applicable for a smaller market which has grown into a huge market in the past few years. Amino acids are not normally excreted by bacteria in substantial amounts as the mechanism of amino acid synthesis is highly controlled. Hence mutants should be generated for over-synthesis of amino acids. Recombinant techniques need to be applied for the production of mutant strains. These techniques help in rapid development of new producers. For example, biotechnological production of L-glutamate requires it to be excreted extracellularly. Absence of L-glutamate in the cells leads to cell death, and hence it is retained by cytoplasmic membrane. There are two ways by which L-glutamate can be excreted outside the cell: presence of a carrier which mediate the excretion of charged L-glutamate without leakage of other ions and a lipid environment of this carrier which triggers its activity. Another factor which favors the production of L-glutamate is the activity of α-ketoglutarate dehydrogenase. Lowering the activity of α-ketoglutarate dehydrogenase prevents conversion of α-ketoglutarate to succinyl-CoA which leads to enhanced conversion of L-glutamate. Likewise, L-threonine can be produced from mutant strains by preventing the formation of

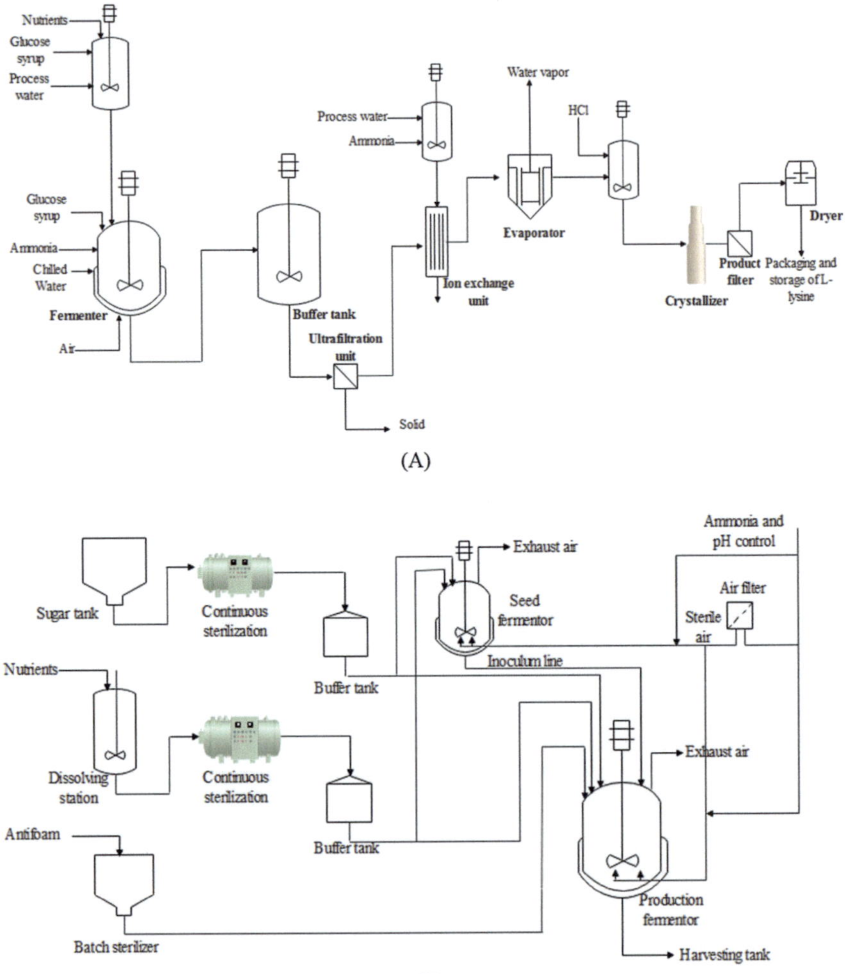

Fig. 3.2 Process layout for the production of l-lysine (**a**) and glutamic acid (**b**). Adapted from Herigemblong (2018) and Garg, respectively

L-isoleucine and increasing the expression of *thrABC* (operon for L-threonine synthesis in *E. coli*). Figure 3.2 depicts the process layout for the production of two amino acids, lysine and glutamic acid.

The production of thermophilic proteins follows three routes. The first method employs thermophilic microorganisms for the production of heat-stable proteins. In the second method, wild-type mesophilic microorganisms that produce

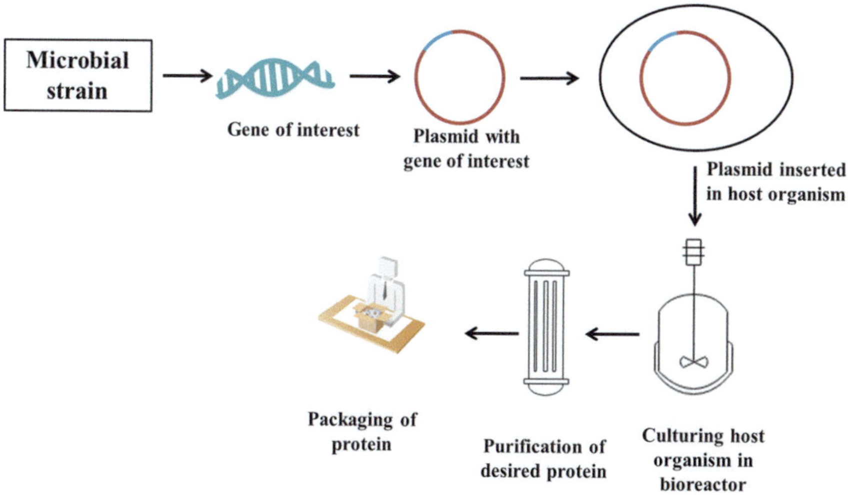

Fig. 3.3 Production, purification, and packaging of recombinant protein. (Adapted from: (GE Reports 2015))

thermostable proteins are explored. And in the third method, recombinant or genetically engineered mesophilic microorganisms can also be created by introducing protein-encoding DNA sequences from thermophiles to produce thermophilic proteins. Figure 3.3 depicts the production and downstream processing of recombinant protein.

Once a suitable microbial strain is selected or modified using recombinant technology, it is cultivated for large-scale production of proteins. The primary consideration in industrial production of proteins is the selection of appropriate fermentation conditions. After the cultivation of producer cells attains a threshold level, the cells are recovered from the medium by centrifugation or filtration. If the product is produced extracellularly by the cells, the fermentation (cell-free) broth is directly used for separation and purification of protein. The separation of intracellular proteins is a tedious task and requires disruption of cells. The cell debris and other macromolecules need to be removed, and the protein of interest is precipitated, filtered, purified using different chromatographic techniques, crystallized, and dried. The dried product is then packaged and shipped (Fig. 3.4a, b).

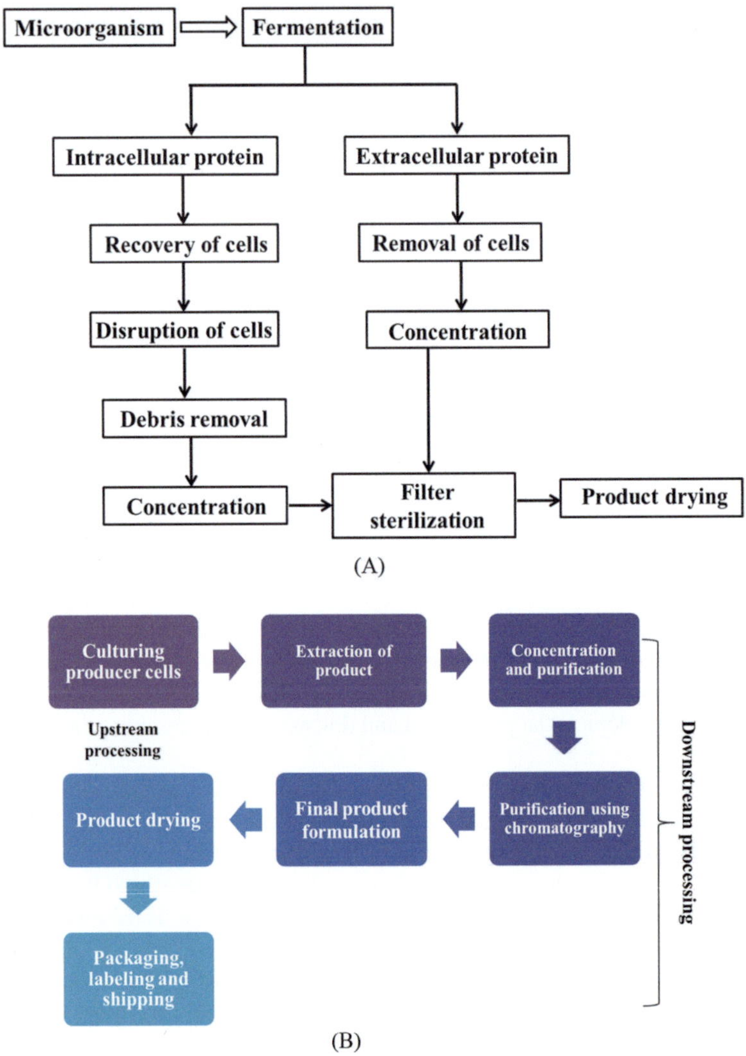

Fig. 3.4 A flowchart for the production (**a**) and downstream processing (**b**) of proteins. (Adapted from (Shuler and Kargi 2002))

References

Arnold FH, Wintrode PL, Miyazaki K, Gershenson A (2001) How enzymes adapt: lessons from directed evolution. Trends Biochem Sci 26(2):100–106

Boutz DR, Cascio D, Whitelegge J, Perry LJ, Yeates TO (2007) Discovery of a thermophilic protein complex stabilized by topologically interlinked chains. J Mol Biol 368(5):1332–1344

Bräsen C, Urbanke C, Schönheit P (2005) A novel octameric AMP-forming acetyl-CoA synthetase from the hyperthermophilic crenarchaeon Pyrobaculum aerophilum. FEBS Lett 579(2):477–482

Cacciapuoti G, Porcelli M, Bertoldo C, De Rosa M, Zappia V (1994) Purification and characterization of extremely thermophilic and thermostable 5′-methylthioadenosine phosphorylase from the archaeon Sulfolobus solfataricus. Purine nucleoside phosphorylase activity and evidence for intersubunit disulfide bonds. J Biol Chem 269(40):24762–24769

Cacciapuoti G, Fuccio F, Petraccone L, Del Vecchio P, Porcelli M (2012) Role of disulfide bonds in conformational stability and folding of 5′-deoxy-5′-methylthioadenosine phosphorylase II from the hyperthermophilic archaeon Sulfolobus solfataricus. Biochim Biophys Acta Proteins Proteomics 1824(10):1136–1143

Cambillau C, Claverie J-M (2000) Structural and genomic correlates of hyperthermostability. J Biol Chem 275(42):32383–32386

Chan C-H, Yu T-H, Wong K-B (2011) Stabilizing salt-bridge enhances protein thermostability by reducing the heat capacity change of unfolding. PLoS One 6(6):e21624

Dahiyat BI, Sarisky CA, Mayo SL (1997) De novo protein design: towards fully automated sequence selection. J Mol Biol 273(4):789–796

de Champdoré M, Staiano M, Rossi M, D'Auria S (2007) Proteins from extremophiles as stable tools for advanced biotechnological applications of high social interest. J R Soc Interface 4(13):183–191

Del Vecchio P, Elias M, Merone L et al (2009) Structural determinants of the high thermal stability of SsoPox from the hyperthermophilic archaeon Sulfolobus solfataricus. Extremophiles 13(3):461–470

Eijsink VG, Gåseidnes S, Borchert TV, van den Burg B (2005) Directed evolution of enzyme stability. Biomol Eng 22(1):21–30

Farias ST, Bonato M (2003) Preferred amino acids and thermostability. Genet Mol Res 2(4):383–393

Fukuchi S, Nishikawa K (2001) Protein surface amino acid compositions distinctively differ between thermophilic and mesophilic bacteria. J Mol Biol 309(4):835–843

Garg M (n.d.) Glutamic acid: history, production and uses (with diagram). Retrieved August 27, 2018, from http://www.biologydiscussion.com/industrial-microbiology-2/glutamic-acid-history-production-and-uses-with-diagram/55763

GE Reports (2015) Recomb production. Retrieved August 31, 2018, from https://www.ge.com/reports/19-technology-breakthroughs-in-2015-that-you-should-know-about/recomb-productiont/

Haney PJ, Badger JH, Buldak GL et al (1999) Thermal adaptation analyzed by comparison of protein sequences from mesophilic and extremely thermophilic Methanococcus species. Proc Natl Acad Sci USA 96(7):3578–3583

Hendsch ZS, Tidor B (1994) Do salt bridges stabilize proteins? A continuum electrostatic analysis. Protein Sci 3(2):211–226

Herigemblong (2018) Fermentation process flow chart awesome industrial production of L lysine by fermentation. Retrieved August 27, 218, from http://dailyrevshare.com/fermentation-process-flow-chart/fermentation-process-flow-chart-awesome-industrial-production-of-l-lysine-by-fermentation/

Ingram-Smith C, Smith KS (2006) AMP-forming acetyl-CoA synthetases in Archaea show unexpected diversity in substrate utilization. Archaea 2(2):95–107

Karshikoff A, Ladenstein R (2001) Ion pairs and the thermotolerance of proteins from hyperthermophiles: a 'traffic rule' for hot roads. Trends Biochem Sci 26(9):550–557
Korkegian A, Black ME, Baker D, Stoddard BL (2005) Computational thermostabilization of an enzyme. Science 308(5723):857–860
Kumar S, Tsai C-J, Nussinov R (2000) Factors enhancing protein thermostability. Protein Eng 13(3):179–191
Laksanalamai P, Maeder DL, Robb FT (2001) Regulation and mechanism of action of the small heat shock protein from the hyperthermophilic Archaeon Pyrococcus furiosus. J Bacteriol 183(17):5198–5202
Lee C-F, Makhatadze GI, Wong K-B (2005) Effects of charge-to-alanine substitutions on the stability of ribosomal protein L30e from Thermococcus celer. Biochemistry 44(51):16817–16825
Luo H, Robb FT (2011) Thermophilic protein folding systems. In: Extremophiles handbook, Springer, Tokyo/Berlin, p 583–599
Malakauskas SM, Mayo SL (1998) Design, structure and stability of a hyperthermophilic protein variant. Nat Struct Mol Biol 5(6):470–475
Mamat B, Roth A, Grimm C et al (2002) Crystal structures and enzymatic properties of three formyltransferases from archaea: environmental adaptation and evolutionary relationship. Protein Sci 11(9):2168–2178
Matthews BW (1993) Structural and genetic analysis of protein stability. Annu Rev Biochem 62(1):139–160
Mayer F, Küper U, Meyer C et al (2012) AMP-forming acetyl coenzyme A synthetase in the outermost membrane of the hyperthermophilic crenarchaeon Ignicoccus hospitalis. J Bacteriol 194(6):1572–1581
Melchionna S, Sinibaldi R, Briganti G (2006) Explanation of the stability of thermophilic proteins based on unique micromorphology. Biophys J 90(11):4204–4212
Park K-H, Kim T-J, Cheong T-K et al (2000) Structure, specificity and function of cyclomaltodextrinase, a multispecific enzyme of the α-amylase family. Biochimt Biophys Acta Protein Struct Mol Enzymol 1478(2):165–185
Park J-T, Song H-N, Jung T-Y et al (2013) A novel domain arrangement in a monomeric cyclodextrin-hydrolyzing enzyme from the hyperthermophile Pyrococcus furiosus. Biochim Biophys Acta Proteins Proteomics 1834(1):380–386
Petsko GA (2001) [34] Structural basis of thermostability in hyperthermophilic proteins, or "There's more than one way to skin a cat". Methods Enzymol 334:469–478
Pühler G, Weinkauf S, Bachmann L et al (1992) Subunit stoichiometry and three-dimensional arrangement in proteasomes from Thermoplasma acidophilum. EMBO J 11(4):1607
Razvi A, Scholtz JM (2006) Lessons in stability from thermophilic proteins. Protein Sci 15(7):1569–1578
Rechsteiner M, Hoffman L, Dubiel W (1993) The multicatalytic and 26 S proteases. J Biol Chem 268:6065–6065
Reed CJ, Lewis H, Trejo E, Winston V Evilia C (2013) Protein adaptations in archaeal extremophiles. Archaea
Ruepp A, Rockel B, Gutsche I, Baumeister W, Lupas AN (2001) The chaperones of the archaeon Thermoplasma acidophilum. J Struct Biol 135(2):126–138
Sadeghi M, Naderi-Manesh H, Zarrabi M, Ranjbar B (2006) Effective factors in thermostability of thermophilic proteins. Biophys Chem 119(3):256–270
Sawle L, Ghosh K (2011) How do thermophilic proteins and proteomes withstand high temperature? Biophys J 101(1):217–227
Shuler ML, Kargi F (2002) Bioprocess engineering: basic concepts, 2nd edn. Prentice Hall, Upper Saddle
Sieber V, Plückthun A, Schmid FX (1998) Selecting proteins with improved stability by a phage-based method. Nat Biotechnol 16(10):955–960

Szilágyi A, Závodszky P (2000) Structural differences between mesophilic, moderately thermophilic and extremely thermophilic protein subunits: results of a comprehensive survey. Structure 8(5):493–504

Taylor TJ, Vaisman II (2010) Discrimination of thermophilic and mesophilic proteins. BMC Struct Biol 10(1):S5

Tomazic SJ, Klibanov AM (1988) Mechanisms of irreversible thermal inactivation of Bacillus alpha-amylases. J Biol Chem 263(7):3086–3091

Unsworth LD, van der Oost J, Koutsopoulos S (2007) Hyperthermophilic enzymes– stability, activity and implementation strategies for high temperature applications. FEBS J 274(16):4044–4056

Vieille C, Zeikus GJ (2001) Hyperthermophilic enzymes: sources, uses, and molecular mechanisms for thermostability. Microbiol Mol Biol Rev 65(1):1–43

Vihinen M (1987) Relationship of protein flexibility to thermostability. Protein Eng Des Sel 1(6):477–480

Vogt G, Argos P (1997) Protein thermal stability: hydrogen bonds or internal packing? Fold Des 2:S40–S46

Vogt G, Woell S, Argos P (1997) Protein thermal stability, hydrogen bonds, and ion pairs. J Mol Biol 269(4):631–643

Wilkinson B, Gilbert HF (2004) Protein disulfide isomerase. Biochim Biophys Acta Proteins Proteomics 1699(1):35–44

Wintrode PL, Arnold FH (2001) Temperature adaptation of enzymes: lessons from laboratory evolution. Adv Protein Chem 55:161–225

Woycechowsky KJ, Raines RT (2003) The CXC motif: a functional mimic of protein disulfide isomerase. Biochemistry 42(18):5387–5394

Zwickl P, Grziwa A, Puehler G et al (1992) Primary structure of the Thermoplasma proteasome and its implications for the structure, function, and evolution of the multicatalytic proteinase. Biochemistry 31(4):964–972

Platform Chemicals by Thermophiles

4

A biologically derived platform chemical originates entirely from biomass and is utilized for the production of various chemical compounds.

1 Organic Acids from Thermophilic Microorganisms

1.1 Butyric Acid

Butyric acid, commonly known as butanoic acid, is a carboxylic acid with a molecular formula of C_3H_7COOH. It is a saturated short-chain fatty acid and is present in milk, butter, cheese, plant oils, and animal fats. It can also be produced in human colon by anaerobic fermentation by bacteria and has been known to suppress colorectal cancer. Butyric acid is used for the production of different low molecular weight butyrate esters, such as methyl butyrate. Due to their pleasant aromas and/or taste, they are used in perfume and food additives. Hence, they are synthesized in industries, chemically or by fermentation. Figures 4.1 and 4.2 show the chemical and biological methods for butyric acid production, respectively. Butyric acid and its derivatives are used as perfumes, additives, drug, diluents, and solvents (Zhang et al. 2009). It is also used in the synthesis of surfactants, plastics, and plasticizers. Current techniques involve the utilization of petrochemicals for the production of butyric acid with the disadvantages of environmental pollution and the exploitation of nonrenewable fuels. To overcome these disadvantages, bio-based methods provide a suitable alternative. These methods are also favorable among consumers as a source of natural ingredients for cosmetics, food, and pharmaceuticals.

A moderate thermophile, *Clostridium thermobutyricum*, with an optimal growth temperature of approximately 55 °C, was used by Wang et al. for bio-based production of butyric acid (Wang et al. 2015). In their work, *Clostridium thermobutyricum* exhibited the ability to ferment sugars present in sweet sorghum juice treated with

J. S. Eswari et al., *Thermophiles for Biotech Industry*,
https://doi.org/10.1007/978-981-32-9919-1_4

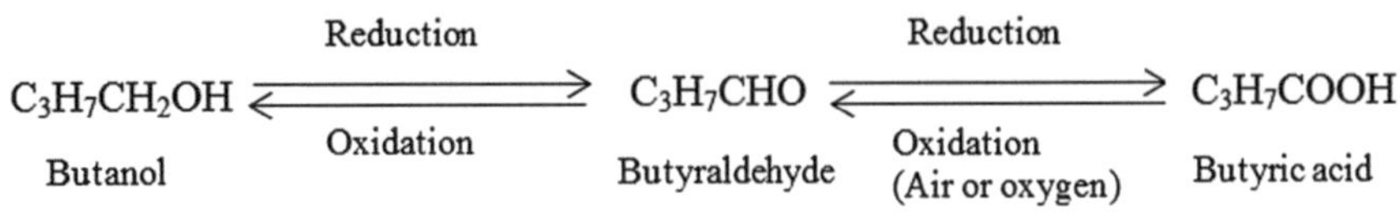

Fig. 4.1 Chemical method of butyric acid production

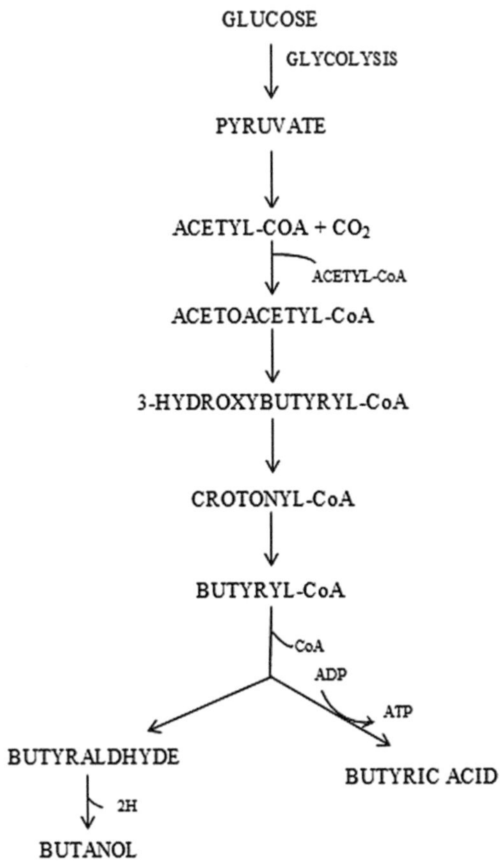

Fig. 4.2 Biological routes of butyric acid production

invertase to 44 g/L butyrate with highest volumetric productivity of 1.45 g/L/h at 50 °C. Along with this, this organism also fermented sugars from acid and enzyme-treated sweet sorghum bagasse to butyrate. The total yield of butyrate from both sweet sorghum juice and bagasse (fresh sweet sorghum stalk) was expected to be 90 kg per tonne of sorghum stalk with a conversion rate of 85% of total sugars. Hence, *Clostridium thermobutyricum* can be used as a microbial alternative for the production of bio-based butyrate from waste as well as renewable feedstock.

1.2 Lactic Acid

Lactic acid was first discovered by a Swedish chemist, Carl Wilhelm Scheele, in 1780. It was isolated from sour milk and hence the name "lactic" which means related to milk. Lactic acid as a fermentation product was demonstrated by a French scientist, Frèmy, and since then the industrial production of lactic acid began. Lactic acid is present in two optically active isomers—L(+) and D(−). Lactic acid, with a molecular formula of $CH_3CH(OH)COOH$, is a hydrocarbon which is white and water-soluble in solid state. Lactate, a conjugate base of lactic acid, has applications in many biochemical processes. Industrially, lactic acid is produced by fermentation of carbohydrates by bacteria (Fig. 4.3) or chemically by formaldehyde (Fig. 4.4). The chemical synthesis produces racemic mixture of lactic acid, whereas bacterial fermentation results in either D(−) or L(+) stereoisomer apart from the racemic mixture of both. The advantages of fermentation of lactic acid than chemical synthesis are the low cost of substrates, high product specificity, and low consumption of energy (Qin et al. 2009).

The lactic acid production by bacteria includes species of *Carnobacterium*, *Enterococcus*, *Lactobacillus*, *Lactococcus*, *Leuconostoc*, *Oenococcus*, *Pediococcus*, *Streptococcus*, *Terragenococcus*, *Vagococcus*, and *Weissella*. The prominent among them are various members of *Lactobacillus* species, namely, *L. acidophilus*, *L. casei*, *L. delbrueckii* subsp. *bulgaricus* (*Lactobacillus bulgaricus*), *L. helveticus*, *Lactococcus lactis*, and *Streptococcus salivarius* subsp. *thermophilus* (*Streptococcus thermophilus*). These bacterial species can be classified into homofermentative and heterofermentative bacterial strains based on the sugar fermentation patterns. Homofermentative bacteria follow Embden-Meyerhof (EM) pathway to ferment hexoses to lactate. Heterofermentative bacteria, on the other hand, utilize pentose phosphate pathway to catabolize sugars to lactate, carbon dioxide, and ethanol in equimolar ratio. Generally, any carbohydrate source with C5 or C6 sugars can be used as a source for the production of lactic acid, but the most commonly used sources are glucose from starch, sugar beet juice, and pure sucrose.

Among the two forms of optically active lactic acid, D-(−) and L-(+), only L-(+) lactic acid is metabolized by human beings and is used to produce plastics. The L-(+) lactic acid is produced by chemical method or biological fermentation. The efficient method for lactic acid production is biosynthesis by homofermentative bacterial strains. Thermophilic bacterial strains for the production of lactic acid provide the advantages of requirement of non-sterile, auto-selective processes with higher productivity and reduction in fermentation cost (Danner et al. 2002). The industrial production of lactic acid requires a series of reactors from the fermentation of the substrate till the evaporation of product (Fig. 4.5).

Lactobacillus helveticus, a homofermentative lactic acid bacterium, produces twice the amount of lactic acid as compared to *Streptococcus thermophilus* and *L. bulgaricus*. A batch fermenter was set up by Roy et al. for the production of a racemic mixture of lactic acid (Roy et al. 1986). The fermenter had a working volume of 400 mL with temperature of 42 °C and agitation rate of 200 rpm. The pH of 5.9

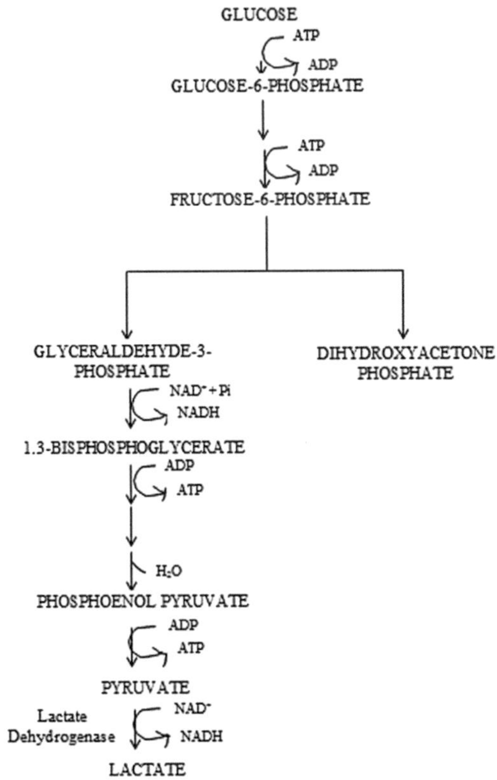

Fig. 4.3 Lactic acid production by biological route

$$CH_3CHO + H_2O + CO \longrightarrow CH_3CH(OH)COOH$$

Acetaldehyde   Water   Carbon monoxide   Lactic acid

Fig. 4.4 Lactic acid production by chemical route

was found to be suitable after optimizing the growth of microorganism and lactic acid production and was maintained by the automatic addition of 3 N ammonia solution. The lactic acid concentration at the end of the fermentation was observed as 35 g/L for duration of 12 h with a substrate conversion rate of 93%. This was less compared to *L. bulgaricus* which produced 60 g/L of lactic acid after 10 h converting 91% of the substrate, but *L. helveticus*, being a thermophile, overcomes the advantage of phage contamination during industrial production.

Bacillus coagulans is a thermophilic species of *Bacillus* that produces lactic acid at 50 °C. *B. coagulans* was used by Payot et al. wherein a two-step method was followed for lactic acid production (Payot et al. 1999). The first step of the operation was a continuous culture to define the aeration and nitrogen requirements. An increase in

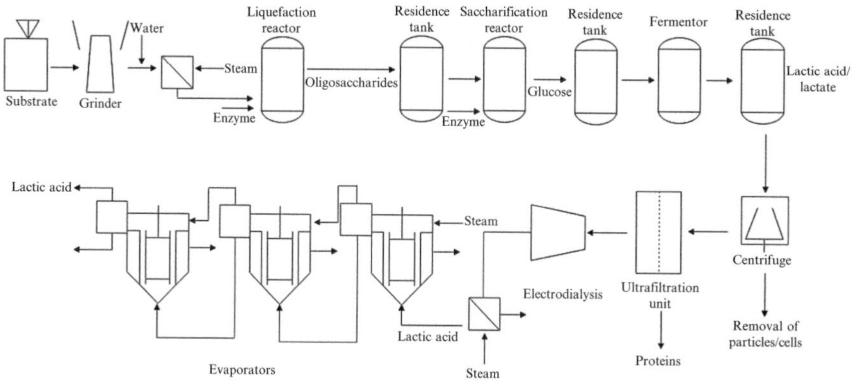

Fig. 4.5 Process layout for lactic acid production

biomass concentration was observed under aerobic condition, whereas lactate production decreased in the same setup. Among the different nitrogen sources tested, yeast extract showed a considerable increase in biomass. This increase was two- and threefold when powdered and liquid yeast extracts were used with the productivities being 13.5 g/L and 19 g/L, respectively. No significant change in biomass concentration was observed when pH was varied from 6.0 to 7.0, but the highest lactic acid production of 22.5 g/L was seen at pH 6.5. Once the experimental parameters, such as aeration, nitrogen source, and pH, were determined, an experimental designed was developed for batch fermentation. During the batch fermentation, the effect of pH, nutrients (sugar and nitrogen), and temperature were optimized. The concentrations of biomass and lactic acid produced were 3.1 ± 0.2 g/L and 55 ± 2.5 g/L, respectively, with maximal specific production rate of lactic acid as 6.1 ± 0.3 /h. A yield of 92 ± 9% of lactate was obtained per initial sucrose concentration. The fermentation conditions optimized for lactic acid production were temperature of 50 or 52 °C, agitation of 300 rpm, and pH of 6.2 or 6.4 maintained by 2–8 N ammonia.

Another thermophilic lactic acid bacterium, *Bacillus* sp. 2-6 isolated from soil, was used to produce polymer-grade L-lactic acid (Qin et al. 2009). Prior to the production of lactic acid by fermentation, the fermentation parameters, namely, temperature, initial glucose concentration, and fermentation medium, were optimized. A temperature range of 40–60 °C was tested for the production of lactic acid, and the highest production was observed to be 66.0 g/L at 50 °C. Different glucose concentrations were tested for lactic acid production. No glucose was consumed when the initial glucose concentration was 254 g/L, whereas the concentrations of 220 g/L and 186 g/L led to limited lactic acid production. The highest amount of lactic acid was produced with 97 g/L and 133 g/L of glucose. The fermentation medium was optimized to consist of glucose, yeast extract, soy peptone, cottonseed protein, sodium nitrate, and ammonium chloride. A 5 liter bioreactor was used for batch fermentation for *Bacillus* sp. 2-6. The lactic acid formation was divided into two phases, 0–15 h and 15–27 h. The lactic acid production reached maximum (118.0 g/L) at the end of 27 h. Pulse feeding strategy was used for fed-batch fermentation. In fed-batch fermentation, the lactic acid production was divided into two phases—0–21 h and 21–60 h. The feed was started at 21 h when the residual glucose concentration reached below 20 g/L. The productivity of lactic acid before 21 h was 100.25 g/L which reached 172.5 g/L at the end of fed-batch fermentation. When scaled-up to 30 liter fed-batch fermentation, the final lactic acid concentration and average productivity were 182.0 g/L and 3.03 g/L/h, respectively.

Liu et al. studied the production of l-(+) lactic acid using thermophilic *Lactobacillus plantarum* As. 1.3 (Liu et al. 2010). They investigated the effects of five nitrogen sources, ammonium chloride (NH_4Cl), ammonium nitrate (NH_4NO_3), corn steep liquor, malt sprout, and diamine citrate. Analysis of results using response surface methodology showed that corn steep liquor and malt sprout affect production of lactic acid significantly with their optimal concentrations in the medium being 12.0 g/L and 16.0 g/L, respectively. The maximum specific growth rate (μ_m) of *Lactobacillus plantarum* As. 1.3 was 1.09 /h with yield coefficient ($Y_{P/S}$) of l-(+) lactic acid as 0.98 g/g. The average volumetric productivity was 3.20 gL^{-1} h^{-1}, and the maximum volumetric productivity was found to be 13.0 gL^{-1} h^{-1}. This study depicted the production of a thermophilic bacterial strain with inexpensive sources of nitrogen.

1.2.1 Applications of Lactic Acid

Lactic acid has a variety of applications in chemical, cosmetic, food, pharmaceutical, and textile industries (Amass et al. 1998; Altıok et al. 2006). Its moisturizing and disinfectant ability makes it useful in cosmetics and pharmaceutical industries. In textile industry, it helps the fabric take up the dyes. Food industry uses lactic acid for the production of yogurt, cheese, pickled food, and sauerkraut. It is used to tan leather and manufacture inks and lacquers. Also, calcium stearoyl-2-lactylate, a compound made from lactic acid, is used as a food preservative. Lactic acid can also be used to derive various products, namely, polylactic acid, polyacrylic acid, acrylate ester, propanoic acid, glycidol, and acetaldehyde (Fig. 4.6). All these compounds have varied applications in pharmaceutical and biomedical sectors.

1.3 Acetic Acid

Acetic acid, systematic name ethanoic acid, is a carboxylic acid with a molecular formula of CH_3COOH. It is a colorless liquid and has varied applications as industrial chemical and chemical reagent. Vinegar, having a 3–9% acetic acid by volume, is one of the applications of acetic acid. Other applications include its use in the production of vinyl acetate, ester, and acetic anhydride and as a polar protic solvent. Apart from its use in industries, it also exhibits antibacterial and antifungal properties against *Enterococci*, *Pseudomonas*, *Staphylococci*, and *Streptococci*. The chemical route of lactic acid production is shown in Fig. 4.7.

Talabardon et al. studied the fermentation of milk permeate for the production of acetic acid under anaerobic and thermophilic conditions (Talabardon et al. 2000). Two thermophilic bacterial strains, *Moorella thermoacetica* DSM 2955 and *M. thermoautotrophica* DSM 741, can convert lactate to acetate at high temperatures

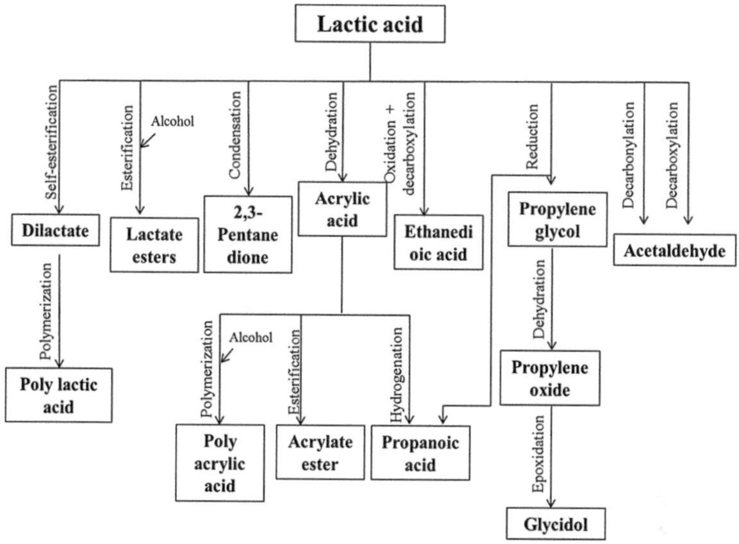

Fig. 4.6 Intermediates in lactic acid metabolism

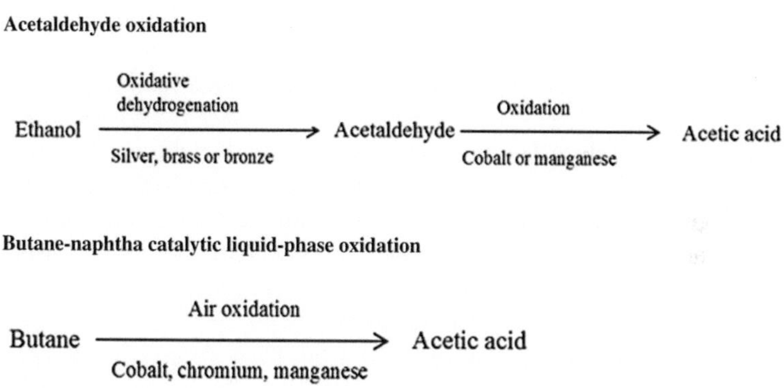

Fig. 4.7 Chemical production of acetic acid

with yields of approximately 0.93 g/g. Another thermophilic bacterium, *Clostridium thermolacticum* DSM 2910, has the ability to produce lactate and acetate from lactose producing large amounts of these compounds. This microorganism also produced substantial amounts of carbon dioxide, hydrogen atoms, and ethanol. The yield of lactate by this organism was dependent on growth of cell. In the exponential phase, the main products of fermentation were ethanol, acetate, carbon dioxide, and hydrogen atom with acetate and ethanol produced in equimolar ratio. The stationary phase of bacteria produced only lactic acid with a yield of 4 mol of lactic acid per mole of lactose. Co-culturing this bacterium with *M. thermoautotrophica* first produces lactic acid from lactose and then acetic acid with an overall acetate yield of 80%. In this process, the yield of ethanol by *C. thermolacticum* from lactose is only 13%.

Another method by which 95% of lactose can be converted to acetic acid is by converting ethanol, produced as a by-product, to acetic acid or converting heterofermentation to homolactic acid fermentation. This can be done by employing a third microorganism which aids in this conversion. Conversion of heterofermentation to homolactic fermentation can be performed by controlling growth phases and fermentation conditions. Moreover, immobilized cell fermentation can be used to reduce growth of cell and increase product yields (Huang and Yang 1998). Talabardon-Albesa suggested that better yield of acetic acid can be obtained from lactose by maintaining co-cultures of *C. thermolacticum* and *M. thermoautotrophica* in stationary phase (Talabardon-Albesa 1999). In stationary phase, the heterofermentative pathway of bacteria shifted to homolactic acid pathway yielding only lactate from lactose with nearly 100% yield. Hence, there is also a possibility of producing acetate from lactose with an approximate yield of 95% using this co-culture.

2 Production of Antimicrobial Agents from Thermophilic Microbial Sources

Any agent which kills microorganisms or retards their growth is termed as antimicrobial agent. They are classified as antibiotics and antifungals if they target bacteria and fungi, respectively. Based on their function, they are further classified as microbicidal that kill microorganisms and biostatic which inhibit the growth of microorganisms.

Many bacteria produce antimicrobial compounds. Antimicrobial resistance is increasing in bacterial species at an alarming rate. Solution to this problem is the development of techniques for discovery of new antimicrobial compounds. *Bacillus* species produce antimicrobial compounds as secondary metabolites which are active against bacteria, viruses, fungi, and protozoa. Production of these antimicrobial compounds is carried out by fermentation.

A thermophilic bacterium was isolated from Thar deserts in Pakistan. The strains, identified as *Bacillus species* SAT4, exhibited antibacterial activity against *Staphylococcus aureus* and *Micrococcus luteus*, whereas *Pseudomonas aeruginosa* showed no effect. The antimicrobial activity of *Bacillus species* SAT4 was only against Gram-positive bacteria. The organism was tested for the production of antimicrobial agents at different fermentation conditions, namely, media, glucose concentrations, nitrogen concentration, pH, temperature, and time of incubation. The maximum antimicrobial activity of antibiotics from *Bacillus species* SAT4 was shown to be for *Staphylococcus aureus* and *Micrococcus luteus* at 55 °C after 48 h of incubation at a pH of 5.0 with 2% glucose. The novel bacterial strain can be used for the production of potential antimicrobial agents at industrial scale (Muhammad et al. 2009).

In another study conducted by Khalil et al., two thermophilic bacteria were isolated from a hot spring in Jordan Valley (Khalil et al. 2006). These two organisms were found to be *Yersinia* sp. 1 and *Aeromonas hydrophila*. The minimum growth temperatures of these organisms were 28 °C and 26 °C, respectively, with optimum growth temperature above 46 °C. *Yersinia* sp. 1 exhibited antimicrobial activity against *Staphylococcus aureus, Escherichia coli, Pseudomonas aeruginosa, Klebsiella pneumonia, Enterobacter* sp. 1, and *Candida albicans*, whereas *Aeromonas hydrophila* had antimicrobial activity against *Staphylococcus aureus, Escherichia coli, Klebsiella pneumonia, Enterobacter* sp. 1, and *Candida albicans* and no activity against *Pseudomonas aeruginosa.*

Different species of *Bacillus* are isolated from hot springs of Lake Bogoria, Kenya. Torome et al. collected 33 microorganisms from a hot spring, out of which phylogenetic analysis classified 9 isolates as members of thermophilic *Bacillus* (Torome et al. 2015). Antibacterial assay of these isolates against various microorganisms depicted their activity against *Escherichia coli* ATCC 25922 and *Pseudomonas aeruginosa* ATCC 27853 with no activity against *S. aureus* ATCC 29213 and *E. coli* 35218.

3 Production of Bioethanol

Ethanol, most popularly known as alcohol or drinking alcohol, is a simple alcohol with a molecular formula of C_2H_5OH. It is a colorless, volatile, and flammable liquid. Ethanol is used in cosmetics and beauty products in astringents, lotions, and hair sprays. As it is miscible with water and organic compounds, it is used as a solvent in paints, lacquers, and varnish. Chemically, it is produced by ethylene with dilute sulfuric acid as a catalyst (Fig. 4.8). The biological route of ethanol production requires glucose as a substrate (Fig. 4.9), and sugarcane is the preferred and cheap substrate for ethanol (Fig. 4.10). Due to the antibacterial activity, it is the main ingredient in hand sanitizers and household cleaning products. In the United States, ethanol has been approved by the Food and Drug Administration (FDA) to be used as a food additive. It is also used as a fuel additive as it has higher octane number than gasoline and reduces air pollution. Conventionally, it is either produced by fermentation of sugars by microorganisms or through petrochemical processes by hydration of ethylene.

Many strains of thermophilic bacteria have shown to produce bacteria using glucose and xylan as substrates (Table 4.1). Different strains of anaerobic *Clostridium thermocellum* ferment cellulose and cellobiose but cannot ferment pentoses to ethanol. On the other hand, strains of *C. thermohydrosulfuricum*, isolated from thermal environments, can ferment cellobiose, pentoses, and hexoses to ethanol. The strain 39E of *C. thermohydrosulfuricum* produces 1.9 mol of ethanol per mole of glucose fermented which is the highest reported ethanol yield (Zeikus et al. 1980).

$$\underset{\text{Ethylene}}{C_2H_4} + H_2O \xrightarrow{\text{Dilute } H_2SO_4 \text{ as catalyst}} \underset{\text{Ethanol}}{C_2H_5OH}$$

Fig. 4.8 Chemical route of ethanol production

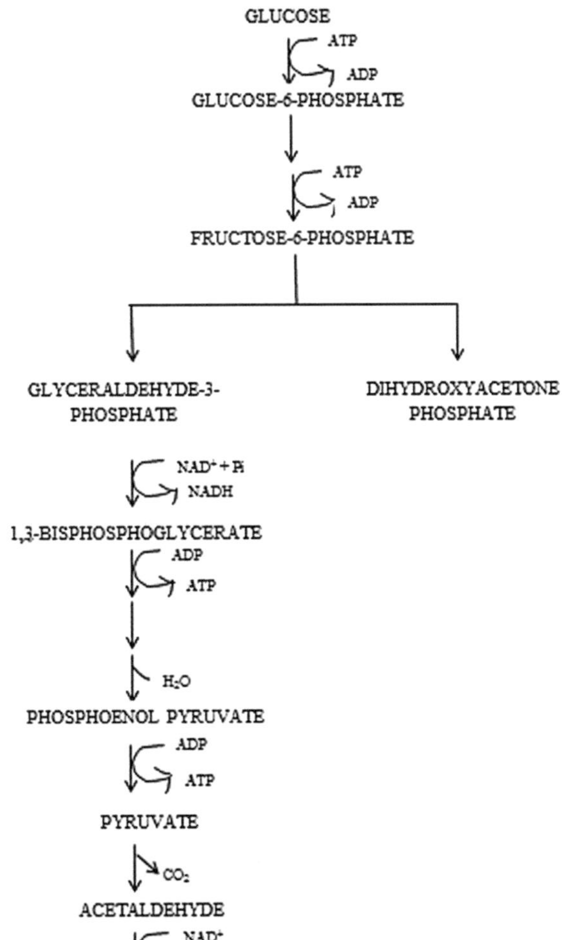

Fig. 4.9 Biological route of ethanol synthesis

Co-cultures of *C. thermocellum* and *C. thermohydrosulfuricum* ferment various cellulosic substrates, including hexoses and pentoses, to produce higher yields of ethanol as compared to monoculture fermentations (Ng et al. 1981).

Some other studies have also considered using microorganisms for ethanol production (Lamed and Zeikus 1980; Lovitt et al. 1988). *C. thermocellum* and *C. thermohydrosulfuricum* are the major organisms that were used. Another genus of thermophilic bacteria capable of producing ethanol is *Thermoanaerobacter*. *T. pentosaceus* is a Gram-negative, rod-shaped bacterium with an optimal growth

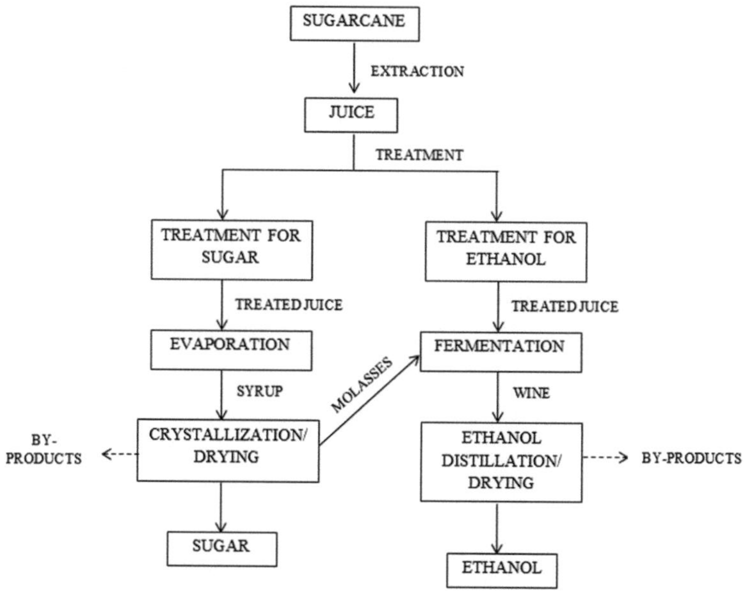

Fig. 4.10 Conventional production of bioethanol by sugarcane

temperature of 70 °C. This organism is capable of fermenting arabinose, cellobiose, fructose, galactose, glucose, lactose, mannitol, mannose, melibiose, pectin, starch, sucrose, xylan, yeast extract, and xylose but cannot ferment acetate, cellulose, glycerol, lactate, rhamnose, lactate, and peptone. This organism produces a maximum yield of 1.39 mol of ethanol per mole of xylose (Tomás et al. 2013b). In order to increase the efficiency of ethanol production, a two-stage process can be used. This method employs the use of *T. pentosaceus* with *Saccharomyces cerevisiae*. In the first step, *S. cerevisiae* utilizes hexoses, and in the second step, *T. pentosaceus* utilizes pentoses to achieve a theoretical yield of 85% ethanol (Tomás et al. 2013a). *Caloramator boliviensis*, a spore-forming thermophilic microorganism, has a growth range of 45–65 °C with optimum temperature being 60 °C. *C. boliviensis* utilizes arabinose, carboxymethylcellulose, cellobiose, fructose, galactose, glucose, glycerol, lactose, maltose, mannose, sucrose, xylan, xylose, and yeast extract as substrates. The highest yield of ethanol produced by this organism was observed to be 0.70 and 0.45 g of ethanol per gram of xylose and cellulose consumed, respectively.

The industrial production of ethanol requires the use of large vessels for processing of substrates and fermentors for the conversion of substrates to products (Fig. 4.11).

Table 4.1 List of thermophilic microorganisms producing ethanol

Microorganism	Max. ethanol yield (mol/mol sugar)		Optimal growth temperature	References
	Glucose	Xylan	(T_{opt}; °C)	
Thermoanaerobacter ethanolicus	1.90	1.64	70	Wiegel and Ljungdahl (1981), Lacis and Lawford (1988), and Hild et al. (2003)
Thermoanaerobacter mathranii	–	1.37	70	Larsen et al. (1997) and Yao and Mikkelsen (2010)
Thermoanaerobacter pentosaceus	–	1.68	70	Tomás et al. (2013a, b)
Thermoanaerobacter pseudoethanolicus	1.88	–	67–69	Zeikus et al. (1980) and Lovitt et al. (1988)
Thermoanaerobacter AK5	1.70	1.35	65	Brynjarsdottir et al. (2012)
Thermoanaerobacter J1	1.70	1.25	65	Jessen and Orlygsson (2012)
Caloramator boliviensis	–	1.53	60	Crespo et al. (2012a, b)
Clostridium thermocellum	1.53	–	60	Rani and Seenayya (1999), Balusu et al. (2004), Tyurin et al. (2004), and Roberts et al. (2010)
Thermoanaerobacterium saccharolyticum	–	1.18	60	Shaw et al. (2008)
Thermoanaerobacterium AK17	1.50	1.33	60	Hild et al. (2003), Koskinen et al. (2008), and Almarsdottir et al. (2012)
Clostridium AK1	1.50	0.85	50	Örlygsson (2012)

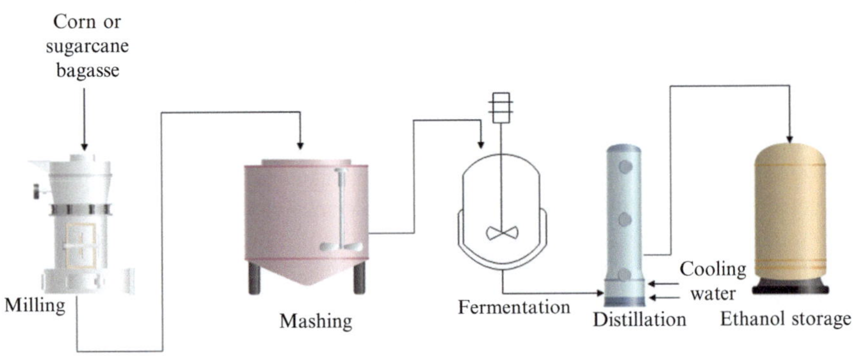

Fig. 4.11 Process layout for the production of bioethanol

4 Growth Kinetics During Batch and Continuous Mode of Operation

4.1 Batch Culture

During bacterial culture, the organisms in seed culture, inoculated in nutrient medium, take up the nutrients and convert them to biomass. The growth of bacteria in batch culture can be differentiated into various phases: lag phase, logarithmic phase, stationary phase, and death phase. The lag phase starts immediately after inoculation. During this phase microorganisms adapt to their new environment and slightly increase their cell mass rather than cell number. The length of this phase depends on the age and volume of inoculum used. The older the inoculum, the lengthier will be the lag phase. The length of this phase also depends on concentrations of some growth factors and nutrients. Figure 4.12 represents the reactor layout of batch process.

In the logarithmic or exponential phase of cell growth, the cells have adopted to the new environment. In this phase, the cells multiply rapidly and the cell mass and cell number increases with time. This is also the phase of balanced cell growth where all components of cell grow at the same rate. During this phase, the specific growth rate can be calculated either from cell mass or cell number which is independent of substrate concentration. The logarithmic growth rate can be represented as:

$$\frac{\mathrm{d}X}{\mathrm{d}t} = \mu_{\mathrm{net}} X, X = X_0 \text{ at } t = 0$$

Integrating this equation yields:

$$\ln\left(\frac{X}{X_0}\right) = \mu_{\mathrm{net}} t$$

or

$$X = X_0 e^{\mu_{\mathrm{net}} t}$$

Fig. 4.12 Batch fermenter

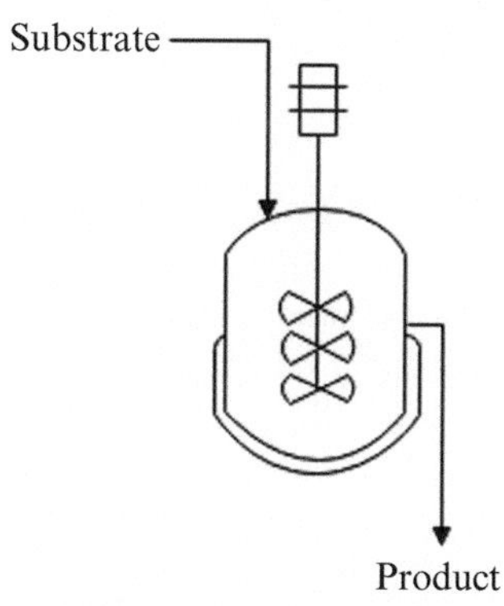

Here X and X_0 are the cell concentrations at time t and $t = 0$.

The time required by the microbial mass to double itself is represented by doubling time:

$$\tau_d = \frac{\ln 2}{\mu_{\text{net}}}$$

or $\tau_d = \dfrac{0.693}{\mu_{\text{net}}}$

During the stationary phase, the overall growth rate is zero or the growth rate is equal to the death rate. The cells are still metabolically active and produce secondary metabolites. In this phase, the cells undergo endogenous metabolism by breaking down cellular reserves for new building blocks. The cells expend energy for essential functions of motility and repair and transport or nutrients. This energy is called maintenance energy. The representation of conversion of cell mass to maintenance energy can be given as:

$$\frac{dX}{dt} = -k_d X$$

or $X = X_{so} e^{-k_d t}$

Here k_d denotes the first-order rate constant for endogenous metabolism, and X_{so} is the cell mass concentration at the beginning of stationary phase.

The death phase begins at the end of stationary phase itself either by exhaustion of essential nutrients or accumulation of toxic products. Lysis of dead cells leads to release of intracellular nutrients in the medium that are used by living cells in stationary phase. The death rate can be depicted as:

$$\frac{dN}{dt} = -k'_d N$$

or $N = N_s e^{-k'_d t}$

Here N_s is the cell concentration at the end of stationary phase, and k'_d is the first-order death rate constant.

4.2 Continuous Culture

In continuous culture, fresh medium is supplied to a culture under continuous agitation, and products and cells are withdrawn simultaneously (Fig. 4.13). Due to continuous cellular growth and product formation, the system reaches a steady state after a certain time, and the cell, substrate, and product concentrations remain constant. Continuous culture helps in determining the response of microorganisms toward their environment and production of desired products under optimal conditions.

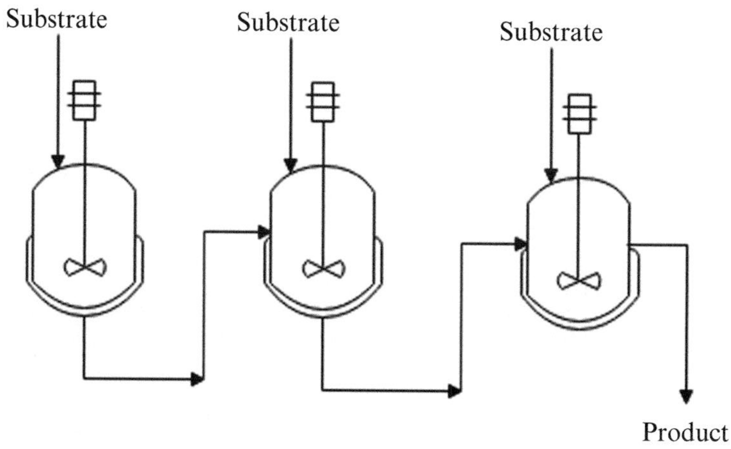

Fig. 4.13 Continuous stirred tank fermenter

References

Almarsdottir AR, Sigurbjornsdottir MA, Orlygsson J (2012) Effect of various factors on ethanol yields from lignocellulosic biomass by Thermoanaerobacterium AK17. Biotechnol Bioeng 109(3):686–694

Altıok D, Tokatlı F, Harsa Ş (2006) Kinetic modelling of lactic acid production from whey by Lactobacillus casei (NRRL B-441). J Chem Technol Biotechnol 81(7):1190–1197

Amass W, Amass A, Tighe B (1998) A review of biodegradable polymers: uses, current developments in the synthesis and characterization of biodegradable polyesters, blends of biodegradable polymers and recent advances in biodegradation studies. Polym Int 47(2):89–144

Balusu R, Paduru RMR, Seenayya G, Reddy G (2004) Production of ethanol from cellulosic biomass by Clostridium thermocellum SS19 in submerged fermentation. Appl Biochem Biotechnol 117(3):133–141

Brynjarsdottir H, Wawiernia B, Orlygsson J (2012) Ethanol production from sugars and complex biomass by Thermoanaerobacter AK5: the effect of electron-scavenging systems on end-product formation. Energy Fuel 26(7):4568–4574

Crespo C, Pozzo T, Karlsson EN, Alvarez MT, Mattiasson B (2012a) Caloramator boliviensis sp. nov., a thermophilic, ethanol-producing bacterium isolated from a hot spring. Int J Syst Evol Microbiol 62(7):1679–1686

Crespo CF, Badshah M, Alvarez MT, Mattiasson B (2012b) Ethanol production by continuous fermentation of D-(+)-cellobiose, D-(+)-xylose and sugarcane bagasse hydrolysate using the thermoanaerobe Caloramator boliviensis. Bioresour Technol 103(1):186–191

Danner H, Madzingaidzo L, Thomasser C, Neureiter M, Braun R (2002) Thermophilic production of lactic acid using integrated membrane bioreactor systems coupled with monopolar electrodialysis. Appl Microbiol Biotechnol 59(2–3):160–169

Hild H, Stuckey D, Leak D (2003) Effect of nutrient limitation on product formation during continuous fermentation of xylose with Thermoanaerobacter ethanolicus JW200 Fe (7). Appl Microbiol Biotechnol 60(6):679–686

Huang Y, Yang ST (1998) Acetate production from whey lactose using co-immobilized cells of homolactic and homoacetic bacteria in a fibrous-bed bioreactor. Biotechnol Bioeng 60(4):498–507

Jessen JE, Orlygsson J (2012) Production of ethanol from sugars and lignocellulosic biomass by Thermoanaerobacter J1 isolated from a hot spring in Iceland. Biomed Res Int 2012:186982

Khalil AB, Zarqa MA, Al-Qaryouti M (2006) Production of antimicrobial agents from thermophilic Yersinia sp. 1 and *Aeromonas hydrophila* isolated from hot spring in Jordan valley. Biotechnology 5(3):252–256

Koskinen PE, Beck SR, Örlygsson J, Puhakka JA (2008) Ethanol and hydrogen production by two thermophilic, anaerobic bacteria isolated from Icelandic geothermal areas. Biotechnol Bioeng 101(4):679–690

Lacis LS, Lawford HG (1988) Ethanol production from xylose by Thermoanaerobacter ethanolicus in batch and continuous culture. Arch Microbiol 150(1):48–55

Lamed R, Zeikus J (1980) Ethanol production by thermophilic bacteria: relationship between fermentation product yields of and catabolic enzyme activities in Clostridium thermocellum and Thermoanaerobium brockii. J Bacteriol 144(2):569–578

Larsen L, Nielsen P, Ahring BK (1997) Thermoanaerobacter mathranii sp. nov., an ethanol-producing, extremely thermophilic anaerobic bacterium from a hot spring in Iceland. Arch Microbiol 168(2):114–119

Liu B, Yang M, Qi B et al (2010) Optimizing l-(+)-lactic acid production by thermophile *Lactobacillus plantarum* As. 1.3 using alternative nitrogen sources with response surface method. Biochem Eng J 52(2):212–219

Lovitt R, Shen G, Zeikus J (1988) Ethanol production by thermophilic bacteria: biochemical basis for ethanol and hydrogen tolerance in Clostridium thermohydrosulfuricum. J Bacteriol 170(6):2809–2815

Muhammad SA, Ahmad S, Hameed A (2009) Antibiotic production by Thermophilic bacillus specie SAT-4. Pak J Pharm Sci 22(3):339–345

Ng TK, Ben-Bassat A, Zeikus J (1981) Ethanol production by thermophilic bacteria: fermentation of cellulosic substrates by cocultures of Clostridium thermocellum and Clostridium thermohydrosulfuricum. Appl Environ Microbiol 41(6):1337–1343

Örlygsson J (2012) Ethanol production from biomass by a moderate thermophile, Clostridium AK1

Payot T, Chemaly Z, Fick M (1999) Lactic acid production by *Bacillus coagulans*—kinetic studies and optimization of culture medium for batch and continuous fermentations. Enzym Microb Technol 24(3–4):191–199

Qin J, Zhao B, Wang X et al (2009) Non-sterilized fermentative production of polymer-grade L-lactic acid by a newly isolated thermophilic strain Bacillus sp. 2–6. PLoS One 4(2):e4359

Rani KS, Seenayya G (1999) High ethanol tolerance of new isolates of Clostridium thermocellum strains SS21 and SS22. World J Microbiol Biotechnol 15(2):173–178

Roberts SB, Gowen CM, Brooks JP, Fong SS (2010) Genome-scale metabolic analysis of Clostridium thermocellum for bioethanol production. BMC Syst Biol 4(1):31

Roy D, Goulet J, LeDuy A (1986) Batch fermentation of whey ultrafiltrate by Lactobacillus helveticus for lactic acid production. Appl Microbiol Biotechnol 24(3):206–213

Shaw AJ, Podkaminer KK, Desai SG et al (2008) Metabolic engineering of a thermophilic bacterium to produce ethanol at high yield. Proc Natl Acad Sci 105(37):13769–13774

Talabardon M, Schwitzguébel J-P, Péringer P (2000) Anaerobic thermophilic fermentation for acetic acid production from milk permeate. J Biotechnol 76(1):83–92

Talabardon-Albesa M (1999) Acetic acid production from milk permeate in anaerobic thermophilic fermentation, Citeseer

Tomás AF, Karagöz P, Karakashev D, Angelidaki I (2013a) Extreme thermophilic ethanol production from rapeseed straw: using the newly isolated Thermoanaerobacter pentosaceus and combining it with *Saccharomyces cerevisiae* in a two-step process. Biotechnol Bioeng 110(6):1574–1582

Tomás AF, Karakashev D, Angelidaki I (2013b) Thermoanaerobacter pentosaceus sp. nov., an anaerobic, extremely thermophilic, high ethanol-yielding bacterium isolated from household waste. Int J Syst Evol Microbiol 63(7):2396–2404

Torome TK, Matasyoh LG, Orinda G, Gakuya F (2015) Isolation and characterization of antibiotic producing Bacillus species in Lake Bogoria, Kenya. Afr J Microbiol Res 9(14):1037–1043

Tyurin MV, Desai SG, Lynd LR (2004) Electrotransformation of Clostridium thermocellum. Appl Environ Microbiol 70(2):883–890

Wang L, Ou MS, Nieves I et al (2015) Fermentation of sweet sorghum derived sugars to butyric acid at high titer and productivity by a moderate thermophile Clostridium thermobutyricum at 50°C. Bioresour Technol 198:533–539

Wiegel J, Ljungdahl LG (1981) Thermoanaerobacter ethanolicus gen. nov., spec. nov., a new, extreme thermophilic, anaerobic bacterium. Arch Microbiol 128(4):343–348

Yao S, Mikkelsen MJ (2010) Metabolic engineering to improve ethanol production in Thermoanaerobacter mathranii. Appl Microbiol Biotechnol 88(1):199–208

Zeikus J, Ben-Bassat A, Hegge P (1980) Microbiology of methanogenesis in thermal, volcanic environments. J Bacteriol 143(1):432–440

Zhang C, Yang H, Yang F, Ma Y (2009) Current progress on butyric acid production by fermentation. Curr Microbiol 59(6):656–663

Biosurfactants, Bioemulsifiers, and Biopolymers from Thermophilic Microorganisms

5

1 Introduction

Secondary metabolites are produced by bacteria, fungi, and plants. They are not directly linked with the normal growth or development of an organism but are associated with functions such as protection and competition. Hence, absence of secondary metabolite will not result in immediate death of organism but will only impair its growth in long run. Mostly, they are specific to species within a phylogenetic group. Microbial secondary metabolites can be classified as antibiotics, anticancer molecules, cholesterol-lowering agents, pigments, etc. These secondary metabolites are generally produced in the stationary phase of bacteria when the macronutrients, namely, carbon, nitrogen, phosphorus, and minerals, have been exhausted in the culture medium. The synthesis machinery (transcription and translation) of secondary metabolites are initiated during nutrient depletion. They have important beneficial effects in humans.

2 Biosurfactants from Thermophiles

Biosurfactants are amphiphilic molecules that consist of both the hydrophilic and hydrophobic parts. They reduce the interfacial and surface tension between molecules at their interface and surface, respectively. They are produced by microorganisms on their cell surface or excreted extracellularly (Karanth et al. 1999). The hydrophilic part of biosurfactant can be an amino acid, a carbohydrate, or a phosphate group, whereas the hydrophobic part is a long chain fatty acid. They can be anionic or neutral. The biosurfactants act by increasing the contact between two immiscible liquids (e.g., oil and water). The two immiscible liquids, when mixed, form separate phases. The addition of biosurfactants leads to dissolution of the hydrophilic part of the biosurfactant in the aqueous phase and the dissolution of hydrophobic part in the lipid phase. This helps in combining the two phases together. Because of this mechanism of action of biosurfactants they play an important role

J. S. Eswari et al., *Thermophiles for Biotech Industry*,
https://doi.org/10.1007/978-981-32-9919-1_5

in treating the hydrocarbon polluted areas. Based on the various hydrophilic and hydrophobic moieties, the biosurfactants can be classified as lipopeptides, glycolipids, phospholipids, polymeric, and particulate biosurfactants. They have a wide range of applications in various industries, for example, for the production of agrochemicals, beverages, cosmetics, emulsifiers, detergents, fertilizers, foaming agents, foods and food ingredients, metallurgy, mining, organic chemicals, petrochemicals, petroleum, pharmaceuticals, and spreading and wetting agents (Vijayakumar and Saravanan 2015).

Chemically, surfactants are produced by petrochemicals by alkylation, ethoxylation, or sulfation to yield different types of surfactant molecules. The major drawbacks of these methods are their polluting and non-renewable nature, and their decomposition leads to release of toxic chemicals. The plant-based surfactants are produced by esterification, hydrogenation, and distillation of plant oils, generally palm and coconut oil. These, being produced from renewable source, have slight advantage over petroleum-based ones. Yet, both the surfactants (petroleum-based and plant-based causes skin allergies and irritation. Biosurfactants, on the other hand, are produced by living cells. The ability of biosurfactants to reduce surface tension is same as that of chemical surfactants. Hence, biosurfactants are preferred for various applications over surfactants as they are renewable and biodegradable and do not have any harmful effect on the consumers. Biosurfactants by marine bacteria have been reported by many researchers for their applications in heavy metal remediation, hydrocarbon degradation, and biomedical applications (Das et al. 2008a, b, 2009; Mukherjee et al. 2009; Sivapathasekaran et al. 2010). The production of biosurfactant by thermophilic microorganisms has several advantages over the mesophilic ones. They have higher reaction rates due to high metabolic activity of the microorganism, reduced risk of contamination by other pathogens due to high incubation temperatures, reduced viscosity of production medium, and reduced cooling costs when using large fermenters.

Thermophilic species of *Bacillus* is a preferred organism for the production of biosurfactant, but some mesophilic *Bacillus* can also produce thermostable biosurfactant. For example, lichenysin from *B. licheniformis* exhibited stability at 50 °C whereas the lipopeptides from *B. subtilis* C9 and *B. subtilis* LB5a were stable even after their incubation at 100° for 1 h and 121 °C for 20 min, respectively (Sen et al. 2012). A species of *B. subtilis* can produce biosurfactant at 45 °C when the medium is supplemented with 2% sucrose and urea or nitrate ions are used as nitrogen source. The biosurfactant reduced the surface tension from 68 mN/m to 28 mN/m and was stable at a wide range of pH (4.5–10.5) and high salinity (4% NaCl) and even after heating at 100 °C for 2 h (Makkar and Cameotra 1997). *B. licheniformis* JF2, a biosurfactant producing microorganism, is capable of growing at temperature as high as 50 °C with pH of 4.6–9 and NaCl concentration of 10%. *B. licheniformis* BAS 50 produces a lipopeptide biosurfactant, lichenysin A, at 35–45 °C and NaCl concentration of 5%, both aerobically and anaerobically. The biosurfactant reduced the surface tension of water from 72 to 28 mN/m. Bioemulsifiers produced by *Methanobacterium thermoautotrophicum* are stable up to 80 °C at a pH range of 5–10 and salt concentrations as high as 200 g/l. These bioemulsifiers do not reduce surface or interfacial tension but form viscous emulsions (Cameotra and Makkar 1998).

Different fermentation regimes can be applied for the production of biosurfactant (Das and Mukherjee 2007) compared to the biosurfactant production efficiency of two *Bacillus subtilis* strains, DM-03 and DM-04 in solid-state fermentation and submerged fermentation systems. In solid-state fermentation, 5 g of potato peels was used as a medium along with 2 ml of M9 medium. Submerged fermentation was carried out in a 5 l Bioflow 110 fermentor with 2% substrate and agitation speed of 200 rpm. The optimum incubation temperatures and pH for the strain DM-03 were 45 °C and 8.0, respectively, and for DM-04 were 55 °C and 7.0, respectively, for both the fermentation systems. The effects of different parameters for biosurfactant production, namely, co-carbon sources, co-nitrogen sources, inoculum size, and initial moisture content of the substrate were studied. Both the strains were capable of reducing the surface tension in the two fermentation systems over the period of 72 h. The maximum yields of biosurfactant produced by DM-03 in solid-state fermentation and submerged fermentations are 67.0 ± 6 mg/gds and 80.0 ± 9 mg/gds, respectively, after 48 h. The strain *B. subtilis* DM-04 produced 20.0 ± 2.5 mg/gds and 23.0 ± 5.0 mg/gds of biosurfactant in solid-state and submerged fermentation systems, respectively.

Another strain of *Bacillus subtilis* 20B was isolated from fermented food and showed antifungal activities against *Alternaria burnsii*, *Chrysosporium indicum*, *Fusarium oxysporum*, *F. udan*, *Rhizoctonia bataticola*, and *Trichoderma herzanium* (Joshi et al. 2008). This organism grows at a temperature of 55 °C with a salt concentration of up to 7%. The organism was shown to utilize alcohols, hydrocarbons, oil, and sugar as carbon source and reduced the surface tension to 29.5 mN/m. A bench scale 5 L fermentor was used for scaling up of biosurfactant production. The conditions of the reaction were 300 rpm, 30 °C, and 1.0 vvm as the aeration rate. The pH of the glucose based minimal medium used was 6.8–7.4. The maximum biosurfactant was produced at 8–10 h of fermentation. Five strains of *Bacillus*, namely, *B. subtilis* R1, *B. licheniformis* R2, *B. licheniformis* K51, *B. subtilis* 20B, and *Bacillus* HS3, were isolated from different sources and used for the production of biosurfactant by Joshi and Desai (2013). A 5-L bench top fermentor equipped with temperature, pH, dissolved oxygen controllers, air sparger, and disc turbine impeller with six blades was used. The batch fermentations were carried out at 30 °C (except for *B. licheniformis* K51 which was 45 °C), 300 rpm with aeration at 1.0 vvm, and pH of 6.8–7.2 maintained with 0.5 N NaOH and 5% o-phosphoric acid. All the bacterial strains were capable of reducing the surface tension to 29–30 mN/m in less than 10 h.

Another class of biosurfactants, glycolipids, is a biomolecule in which carbohydrates are attached to the lipids via glycosidic bonds. Their major role of glycolipids is in the maintenance of cell membrane stability and cellular recognition. There are three major types of glycolipids: rhamnolipids, sophorolipids, and mannosylerythritol lipids. Rhamnolipids consist of β-hydroxy-fatty acid (hydrophobic moiety) which is linked to a rhamnose sugar molecule (hydrophilic moiety) at its carboxyl terminal. The rhamnose sugars are connected to each other via 1,2-glycosidic bonds. The crystalline acid is produced by *Pseudomonas aeruginosa* and can be categorized into mono- and dirhamnolipids. Other bacterial species, namely, *P. chlororaphis*, *P. fluorescens*, *P. plantarii*, and *P. putida*, *Acinetobacter calcoaceticus*, *Burkholderia* sp., *Enterobacter* sp., *Pantoea* sp., and *Pseudoxanthomonas* sp., also produce rhamnolipids (Řezanka et al. 2011; Sekhon Randhawa and Rahman 2014). The rhamnolipids are used in agriculture, environment, and industries as antifungal, antibacterial, and antiviral agents and wetting agents and for metal sequestration and enhanced oil recovery (Costa et al. 2010). Řezanka et al. exploited three thermophilic strains, *Meiothermus ruber* CCM4212, *Thermus* sp. CCM 2842, and *Thermus aquaticus* CCM 3488, for rhamnolipid production. The organisms produce mono- and dirhamnolipids with one or two 3-hydroxy-fatty acids containing 24 carbon atoms. The structure of rhamnolipid was different from that produced by *Pseudomomas aeruginosa*. The advantages of using these organisms for rhamnolipid production is that these organisms fall into class 1 biosafety level, not pathogenic to humans and have better surfactant properties than *P. aeruginosa* (Řezanka et al. 2011).

Sophorolipids are produced by non-pathogenic strains of yeasts. It consists of aglycone as the fatty acid and sophorose as the sugar residue. Sophorose is a disaccharide which consists of two glucose unit linked via β-1,2 glycosidic bond. Sophorolipids exist as monomeric and dimeric forms of lactone wherein the carboxylic group of fatty acid is esterified internally (Kulakovskaya and Kulakovskaya 2013). Figure 5.1 shows the biosynthetic pathway for sophorolipid. Sophorolipids have known to be produced by *Candida apicola*, *C. batistae*, *C. bombicola*, *C. magnolia*, *Rhodotorula bogoriensis*, and *Wickerhamiella domercqiae* (Kurtzman et al. 2010). *Candida bombicola* ATCC 22214 has been shown to produce up to 400 g/l of sophorolipids (Van Bogaert et al. 2011). A thermotolerant strain of yeast, *Pichia anomala* PY1, isolated from fermented foods was shown to produce sophorolipid in a medium containing 4% soybean oil as the sole carbon source. The culture conditions were maintained at 30 °C, pH 5.5 for 7 days, and the sophorolipid, thus produced, reduced the surface tension to 28 mN/m. This organism produces sophorolipid with C20 and C18:1 fatty acid chains in media containing glucose and soybean oil as the sole carbon sources (Thaniyavarn et al. 2008).

Mannosylerythritol lipids (MELs) are surfactants belonging to glycolipid class of biosurfactants. They consist of 4-O-β-D-mannopyranosyl-meso-erythritol as the hydrophilic group and a fatty acid and/or acetyl group as hydrophobic group. They are produced by *Pseudozyma* sp. as the chief component. Other than *Pseudozyma*, MELs are also produced by *Ustilago* sp. and *Schizonella melanogramma*. MELs are produced at the stationary stage of yeast cells and are storage components of yeast cells. They have the ability to reduce the surface tension of water to less than 30 mN/m. MELs have various pharmaceutical applications, such as antitumor activity, treatment of microbial infections, diseases caused by dopamine dysfunction, and schizophrenia. They also exhibit different induced activity against many cancer cell lines. They have the property to self-assemble which can be utilized in drug delivery and gene transfection experiments (Arutchelvi et al. 2008). The yeast, *Pseudozyma parantarctica* JCM 11752, can grow above 40 °C, and hence they can be called as thermotolerant yeasts. This is the only thermostable yeast species capable of producing mannosylerythritol lipids (MELs). This organism produces 100 g/l of diacylated MELs at 34 °C using soybean oil as the carbon source and sodium nitrate as

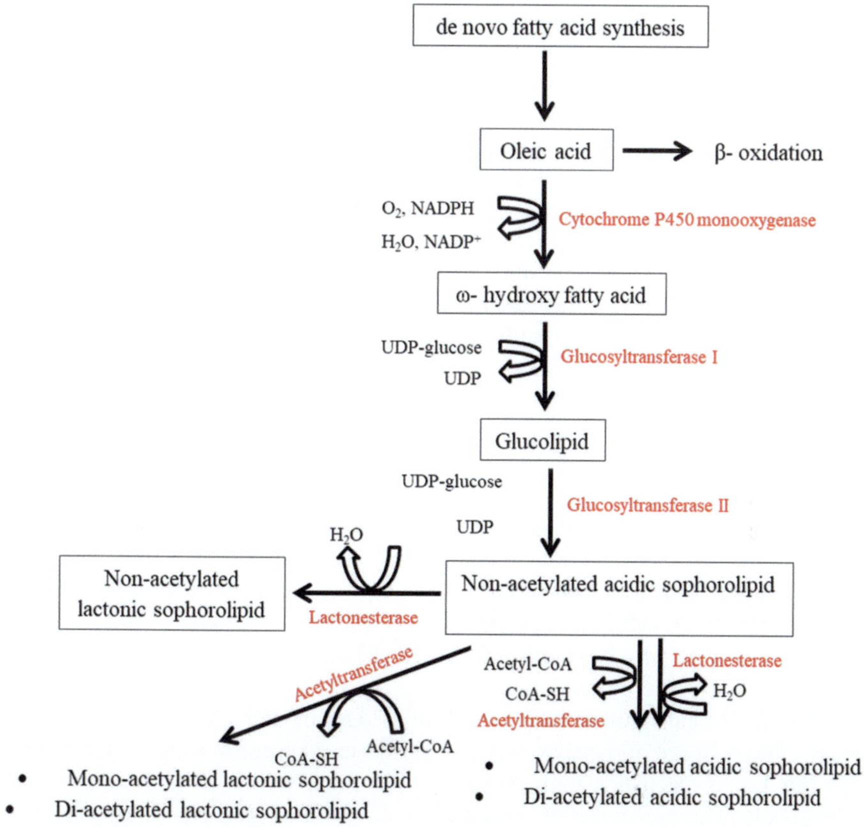

Fig. 5.1 Pathway for sophorolipid biosynthesis. (Adopted from: Van Bogaert et al. 2011; Saerens et al. 2015)

the nitrogen source. Increasing the concentrations of soybean oil leads to the production of di- and triacylated MELs. The maximum production of triacylated MELs was 22.7 g/l at 34 °C with 20% soybean oil (Morita et al. 2008).

3 Bioemulsifiers Production from Thermophiles

Bioemulsifiers are high molecular weight amphiphilic compounds and consist of proteins, lipoproteins, lipopolysaccharides, and heteropolysaccharides. These biomolecules can emulsify two immiscible phases but are inefficient to reduce surface tension between these two phases. They increase the solubility of insoluble or poorly soluble substrates. They stabilize emulsions by increasing the kinetic stability which is because of large number of reactive groups present in their structures. At the interface of two phases, the hydrophobic part of the bioemulsifier is present at the surface in contact with the hydrophobic phase, whereas the hydrophilic moiety is oriented toward the hydrophilic phase. Because of their varied properties like cleansing, foaming, wetting, emulsification, and reduction in viscosity, they find applications in pharmaceuticals, food, cosmetics, and bioremediation processes (Gautam and Tyagi 2006; Franzetti et al. 2010; Perfumo et al. 2010; Fracchia et al. 2012; Uzoigwe et al. 2015).

One of the first reports of bioemulsifier production by a thermophilic microorganism was by Gurjar et al. *Bacillus stearothermophilus* VR-8 was isolated from hot spring and was grown in crude oil containing medium for bioemulsifier production. Though the organism could produce bioemulsifier over the temperature range of 45–70 °C, the maximum production was seen at 50 °C. The bioemulsifier consisted of protein, carbohydrate, and lipids in the concentrations of 46%, 16%, and 10%, respectively. The biomolecule had a broad stability of pH (2–8), temperature (50–80 °C), and salt concentrations (1% $MgCl_2$, 5% $CaCl_2$, and 5% NaCl) (Gurjar et al. 1995). A bacterium, *Methanobacterium thermoautotropicum*, was shown to produce cell-associated bioemulsifier. The bioemulsifier was tested for stability at different temperature, pH and NaCl concentrations and was found to be stable up to 80 °C and 200 g/l salinity with a molecular weight of more than 5000 Da. The bioemulsifier was capable of forming stable emulsions without reducing surface tension of water. As these emulsifiers can tolerate high salt concentrations, they find suitable applications as agents to control mobility in oil or saline reservoirs and in cleanup process of well-bores (de Acevedo and McInerney 1996).

Geobacillus pallidus XS2 and XS3, isolated from soil samples in an oil field in China, was shown to produce bioemulsifiers by utilizing different hydrocarbons. The two strains gave bioemulsifier yields of 4.24±0.08 and 3.82±0.11 g/l, respectively. The average molecular weights of these molecules were 271,785 and 526,369 Da, respectively, with polydispersity index values of 1.104 and 1.027, respectively. The bioemulsifier produced by *G. pallidus* XS2 consisted of 68.6% carbohydrates, 22.7% lipids, and 8.7% proteins, whereas bioemulsifier produced by XS3 had 41.1% carbohydrates, 47.6% lipids, and 11.3% proteins. These emulsifiers were capable of degrading approximately 250 mg/l of polyaromatic hydrocarbons and 500 mg/l of crude oil after 20 days and were stable over a wide range of pH, temperature, and salinity (Zheng et al. 2011). *Anoxybacillus* sp. WJ4, isolated from oil reservoir, produced bioemulsifier at 65 °C. *Paenibacillus* sp. #510 is capable of producing bioemulsifier at 40 °C under aerobic and anaerobic conditions. The emulsions formed by the bioemulsifier was stable and the emulsification ability was independent of its exposure to high temperature (100–121 °C), high pH (2–13), and high salinities (up to 300 g/l) (Gudiña et al. 2015). Table 5.1 summarizes the bioemulsifier production of few of the thermophilic bacterial species.

4 Biopolymers from Thermophilic Microorganisms

Biopolymers are high molecular weight polymers produced by living organisms, for example, microorganisms and plants. They are macromolecules of single repeating monomers with varied monomer composition. They are produced from renewable and carbon-negative feedstock and hence are suitable alternatives for petroleum

Table 5.1 Bioemulsifier production temperature and source of isolation of thermophilic bacteria

S. No.	Microorganism	Source of thermophilic bacteria	Bioemulsifier production: optimal temperature for growth and production of bioemulsifier (°C)	References
1.	*Bacillus stearothermophilus* VR-8	Hot spring	50	Gurjar et al. (1995)
2.	*Methanobacterium thermoautotropicum*	Procured	60	de Acevedo and McInerney (1996)
3.	*Geobacillus pallidus* XS2 and XS3	Soil samples in oil field	60	Zheng et al. (2011)
4.	*Anoxybacillus* sp. WJ-4	Oil samples from reservoir	65	Xia et al. (2015)
5.	*Paenibacillus* sp. #510	Crude oil	40	Gudiña et al. (2015)

derived synthetic polymers. Due to their biocompatibility and ease of processing, they are used in medicine, agriculture, chemical, and packaging industries. Polyhydroxyalkanoates (PHAs) are polymers of 3-,4-,5- and 6-hydroxyalkanoic acid. They are synthesized by microorganisms under nutrient-limiting condition and in the presence of excess carbon source. They are insoluble in water, biodegradable, and biocompatible and hence are used in biomedical applications.

In one study, six bacterial strains were isolated from hot springs in Egypt and waste treatment plant in Germany and were found to be closely related to *Chelatococcus* species. All the strains exhibited optimum growth at 50 °C. Four of the strains produced poly(3-hydroxybutyrate) in the presence of glucose, and the remaining two utilized glycerol for PHB production. The highest PHB accumulation of 73% (w/w) was observed at 50 °C with the dry cell density of 4.8 g/l (Ibrahim et al. 2010). In another study, *C. daeguensis* TAD1 was isolated from biofilm of a trickling filter and used for PHB production. The results of the study showed that PHB accumulation in the organism is growth-associated and does not require any nutrient-limiting condition. *C. daeguensis* TAD1 produced 83.6% (cell dry weight) of PHB at 45 °C in 24 hours whereas the maximum amount of PHB (3.44±0.3 g/l) was shown at 50 °C when glucose was used as the carbon source. Also, the organism is capable of utilizing starch or glycerol from cheap substrates as raw material for the production of PHB which infers that the organism can be successfully employed for the large scale production of PHB from waste substrates, particularly those that are rich in glycerol or starch (Xu et al. 2014).

Aneurinibacillus sp. XH2 was isolated from an oil field in China and used for PHA production. The medium for PHA production consisted of only yeast extract, peptone, and glucose. The yield of PHA in this medium was 111.6 mg/l at 55 °C which increased to 2.4 times when peptone was removed. The monomers of this PHA were 3-hydroxyvalerate and 3-hydroxybutyrate, as identified by GC-MS (gas chromatography-mass spectroscopy) (Xiao et al. 2015). Giedraitytė and Kalėdienė isolated *Geobacillus* sp. AY 946034 from an oil field in Lithuania. MSM containing 1% glucose as the sole carbon source was used for production of poly-3-hydroxybutyrate at temperature of 60 °C. The highest production of PHB was 68.9% of cell dry weight and was characterized using various analytical techniques. The thermal degradation temperature of the PHB, thus produced, was 280 °C with melting temperature of 168.8 °C and molecular weight of 556 kDa (Giedraitytė and Kalėdienė 2015).

Exopolysaccharides are also high molecular weight polymers produced by microorganisms that are secreted into the surrounding. They are classified as homopolysaccharides (identical sugar residues) and heteropolysaccharides (different sugar residues). Four new polysaccharides were isolated from several thermophilic bacteria from shallow hydrothermal vents and marine hot springs, near Lucrino area and around Ischia Island, Italy. They were capable of producing polysaccharides at 65 °C. A strain, *Bacillus thermantarcticus*, produced two extracellular polysaccharides with the production of 400 mg/l. The exopolysaccharides were found to be composed of four different α-D-mannoses with three different β-D-glucoses and mannan, respectively (Nicolaus et al. 2004).

5 Bioprocess Considerations for the Production of Biosurfactants and Biopolymers

Lipopeptides, produced as secondary metabolites by many bacterial species, have various pharmaceutical applications as antibacterial, antifungal, antiviral, and antitumor agents. These are the good candidates to replace those drugs which are already resistant in some strains of microorganisms. The major drawbacks of production and use of these molecules are their high production cost and low yield. The production of lipopeptides can be increased by the use of different carbon sources and optimum culture conditions. For example, wastes such as potato peels, rice bran, soybean, and molasses have been used as cheap raw materials for lipopeptide production. The lipopeptide yield can be further increased by the addition of divalent cations, preferably Fe^{2+} and Mn^{2+} (Wei and Chu 2002; Rangarajan et al. 2012). Lipopeptides are growth-associated molecules and are produced when one or more limiting nutrients have been depleted. Fengycin are produced in the early stationary phase, iturins in later stationary phase whereas surfactins are synthesized during the transition of cells from exponential to stationary phase (Ongena and Jacques 2008).

Carbon source is an important factor as it considers the cost for the production of lipopeptides. The nature of carbon source directly depends upon the structure and composition of lipopeptides. Surfactin is produced when dextrose is used as the carbon source whereas fengycin and iturin are produced in the presence of glycerol and sucrose. The antifungal ability of lipopeptides is also affected by the substrate used as the bacteria grown in dextrose or sucrose exhibits higher antifungal activity (Singh et al. 2014). Inorganic compounds used for lipopeptide production are ammonium nitrate, ammonium sulfate, monosodium glutamate, sodium nitrate, and urea and cheap raw materials such as casein hydrolysate, peptone, and soybean flour. Tryptone is the most favorable organic nitrogen for lipopeptide production as it contains many amino acids that are homologous to lipopeptides (Zhao et al. 2013). The culture conditions, such as temperature, pH, concentration of dissolved oxygen, and aeration rate also affect the lipopeptide production. These conditions are primarily dependent on the microbial strain used. The oxygen transfer is an important parameter for the production of surfactin as changes in oxygen transfer might result in either monoproduction of surfactin or a mixed population of lipopeptides (Chen et al. 2015). Foaming presents an obstruction in the transfer of oxygen as it decreases the rate of oxygen transfer.

The design of bioreactor is another factor to be considered in any industrial bioprocess. The choice of appropriate bioreactor design helps in increasing the productivity of any biological product. Lipopeptides are produced by liquid fermentation or solid-state fermentation. In liquid fermentation, the application of bioreactors with gas bubbling is not generally preferred for lipopeptide production as they generate excessive foam. The antifoam agents affect the bacterial growth and increase the purification step. Foam can be controlled by the use of anaerobic reactor for lipopeptide production. Other alternatives are the use of bubbleless membrane aerated bioreactor, rotating discs biofilm reactor, three-phase inverse fluidized bed bioreactor, and foam collector (Chtioui et al. 2012; Coutte et al. 2013; Fahim et al. 2013). On the other hand, solid-state fermentation utilizes agrowastes as substrates which reduces the cost of production. The factors to be considered in solid-state fermentation are the substrates used, fermentation time, and temperature, initial moisture content and the amount of mineral salts (Zhu et al. 2013).

References

Arutchelvi JI et al (2008) Mannosylerythritol lipids: a review. J Ind Microbiol Biotechnol 35(12):1559–1570

Cameotra SS, Makkar R (1998) Synthesis of biosurfactants in extreme conditions. Appl Microbiol Biotechnol 50(5):520–529

Chen W-C et al (2015) Applications of a lipopeptide biosurfactant, surfactin, produced by microorganisms. Biochem Eng J 103:158–169

Chtioui O et al (2012) Rotating discs bioreactor, a new tool for lipopeptides production. Process Biochem 47(12):2020–2024

Costa SG et al (2010) Structure, properties and applications of rhamnolipids produced by *Pseudomonas aeruginosa* L2-1 from cassava wastewater. Process Biochem 45(9):1511–1516

Coutte F et al (2013) New integrated bioprocess for the continuous production, extraction and purification of lipopeptides produced by Bacillus subtilis in membrane bioreactor. Process Biochem 48(1):25–32

Das K, Mukherjee AK (2007) Comparison of lipopeptide biosurfactants production by Bacillus subtilis strains in submerged and solid state fermentation systems using a cheap carbon source: some industrial applications of biosurfactants. Process Biochem 42(8):1191–1199

Das P et al (2008a) Antimicrobial potential of a lipopeptide biosurfactant derived from a marine Bacillus circulans. J Appl Microbiol 104(6):1675–1684

Das P et al (2008b) Improved bioavailability and biodegradation of a model polyaromatic hydrocarbon by a biosurfactant producing bacterium of marine origin. Chemosphere 72(9):1229–1234

Das P et al (2009) Biosurfactant of marine origin exhibiting heavy metal remediation properties. Bioresour Technol 100(20):4887–4890
de Acevedo GT, McInerney MJ (1996) Emulsifying activity in thermophilic and extremely thermophilic microorganisms. J Ind Microbiol 16(1):1–7
Fahim S et al (2013) Oxygen transfer in three phase inverse fluidized bed bioreactor during biosurfactant production by Bacillus subtilis. Biochem Eng J 76:70–76
Fracchia L et al (2012) Biosurfactants and bioemulsifiers biomedical and related applications–present status and future potentials. Biomed Sci Eng Technol, InTech
Franzetti A et al (2010) Applications of biological surface active compounds in remediation technologies. In: Biosurfactants. Springer, New York, pp 121–134
Gautam K, Tyagi V (2006) Microbial surfactants: a review. J Oleo Sci 55(4):155–166
Giedraitytė G, Kalėdienė L (2015) Purification and characterization of polyhydroxybutyrate produced from thermophilic Geobacillus sp. AY 946034 strain. Chemija 26(1):38
Gudiña EJ et al (2015) Novel bioemulsifier produced by a Paenibacillus strain isolated from crude oil. Microb Cell Factories 14(1):14
Gurjar M et al (1995) Bioemulsifier production by Bacillus stearothermophilus VR-8 isolate. Lett Appl Microbiol 21(2):83–86
Ibrahim M et al (2010) Isolation and characterization of new poly (3HB)-accumulating star-shaped cell-aggregates-forming thermophilic bacteria. J Appl Microbiol 109(5):1579–1590
Joshi S et al (2008) Production of biosurfactant and antifungal compound by fermented food isolate Bacillus subtilis 20B. Bioresour Technol 99(11):4603–4608
Joshi SJ, Desai AJ (2013) Bench-scale production of biosurfactants and their potential in ex-situ MEOR application. Soil Sediment Contam Int J 22(6):701–715
Karanth N et al (1999) Microbial production of biosurfactants and their importance. Curr Sci 77:116–126
Kulakovskaya E, Kulakovskaya T (2013) Structure and occurrence of yeast extracellular glycolipids. Extracellular glycolipids of yeasts: biodiversity, biochemistry, and prospects, Academic, pp 1–14
Kurtzman CP et al (2010) Production of sophorolipid biosurfactants by multiple species of the Starmerella (Candida) bombicola yeast clade. FEMS Microbiol Lett 311(2):140–146
Makkar R, Cameotra SS (1997) Biosurfactant production by a thermophilic Bacillus subtilis strain. J Ind Microbiol Biotechnol 18(1):37–42
Morita T et al (2008) Efficient production of di-and tri-acylated mannosylerythritol lipids as glycolipid biosurfactants by Pseudozyma parantarctica JCM 11752T. J Oleo Sci 57(10):557–565
Mukherjee S et al (2009) Antimicrobial biosurfactants from marine Bacillus circulans: extracellular synthesis and purification. Lett Appl Microbiol 48(3):281–288
Nicolaus B et al (2004) Polysaccharides from extremophilic microorganisms. Orig Life Evol Biosph 34(1–2):159–169
Ongena M, Jacques P (2008) Bacillus lipopeptides: versatile weapons for plant disease biocontrol. Trends Microbiol 16(3):115–125
Perfumo A et al (2010) Production and roles of biosurfactants and bioemulsifiers in accessing hydrophobic substrates. In: Handbook of hydrocarbon and lipid microbiology, Springer, pp 1501–1512
Rangarajan V et al (2012) Time-dependent dosing of Fe2+ for improved lipopeptide production by marine Bacillus megaterium. J Chem Technol Biotechnol 87(12):1661–1669
Řezanka T et al (2011) Rhamnolipid-producing thermophilic bacteria of species Thermus and Meiothermus. Extremophiles 15(6):697
Saerens KM et al (2015) Characterization of sophorolipid biosynthetic enzymes from Starmerella bombicola. FEMS Yeast Res 15(7):fov075
Sekhon Randhawa KK, Rahman PK (2014) Rhamnolipid biosurfactants—past, present, and future scenario of global market. Front Microbiol 5:454
Sen R et al (2012) Biosurfactants: synthesis, properties and applications in environmental bioremediation. In: Bioremediation and sustainability: research and applications, pp 137–211

Singh AK et al (2014) Substrate dependent in vitro antifungal activity of Bacillus sp strain AR2. Microb Cell Factories 13(1):67

Sivapathasekaran C et al (2010) Artificial neural network modeling and genetic algorithm based medium optimization for the improved production of marine biosurfactant. Bioresour Technol 101(8):2884–2887

Thaniyavarn J et al (2008) Production of sophorolipid biosurfactant by Pichia anomala. Biosci Biotechnol Biochem 72(8):2061–2068

Uzoigwe C et al (2015) Bioemulsifiers are not biosurfactants and require different screening approaches. Front Microbiol 6:245

Van Bogaert IN et al (2011) Microbial synthesis of sophorolipids. Process Biochem 46(4):821–833

Vijayakumar S, Saravanan V (2015) Biosurfactants-types, sources and applications. Res J Microbiol 10(5):181

Wei Y-H, Chu I-M (2002) Mn 2+ improves surfactin production by Bacillus subtilis. Biotechnol Lett 24(6):479–482

Xia W et al (2015) Hydrocarbon degradation by a newly isolated thermophilic Anoxybacillus sp. with bioemulsifier production and new alkB genes. RSC Adv 5(124):102367–102377

Xiao Z et al (2015) Thermophilic production of polyhydroxyalkanoates by a novel Aneurinibacillus strain isolated from Gudao oilfield, China. J Basic Microbiol 55(9):1125–1133

Xu F et al (2014) Comparative study on the production of poly (3-hydroxybutyrate) by thermophilic Chelatococcus daeguensis TAD1: a good candidate for large-scale production. Appl Microbiol Biotechnol 98(9):3965–3974

Zhao P et al (2013) Effects of critical medium components on the production of antifungal lipopeptides from *Bacillus amyloliquefaciens* Q-426 exhibiting excellent biosurfactant properties. World J Microbiol Biotechnol 29(3):401–409

Zheng C et al (2011) Hydrocarbon degradation and bioemulsifier production by thermophilic Geobacillus pallidus strains. Bioresour Technol 102(19):9155–9161

Zhu Z et al (2013) Enhancement of lipopeptides production in a two-temperature-stage process under SSF conditions and its bioprocess in the fermenter. Bioresour Technol 127:209–215

Applications of Thermophiles

6

1 Thermophiles as Key Players in Environmental Bioremediation

Bioremediation, in environmental biotechnology, is a strategy to control pollution naturally with the help of biological species. The biological species (microorganisms, plants, fungi, or their enzymes) catalyze the biodegradation or biotransformation of toxic chemicals to less toxic forms and hence return the contaminated environment to its original state. EPA describes bioremediation as "treatment that uses naturally occurring organisms to break down hazardous substances into less toxic or non-toxic substances." The conventional techniques of environment remediation require high input of power and water for degradation of pollutants by thermally enhanced recovery, thermal incineration and desorption, chemical treatment, and in situ flushing of soil by water which further requires treatment of flushed water. The microorganisms cannot degrade inorganic contaminants, and hence not all microorganisms can be used for bioremediation. They can change the valence state of inorganic compounds, precipitate, adsorb, or accumulate the contaminants intracellularly, thereby decreasing the concentration of these inorganic contaminants.

Bioremediation can be classified into intrinsic where bioremediation occurs naturally or biostimulation where the process is enhanced by the addition of certain substances, for example, oxygen or fertilizers that promote the growth of pollutant-degrading microorganisms. The three main components for bioremediation are the pollutant or contaminant, microorganisms capable of degrading the pollutant, and an electron acceptor. Generally, the bioremediation process is improved under nutrient-limiting conditions as these compounds are considered as a form of nutrient for the organism. The degradation of synthetic chemicals or xenobiotics requires the synthesis of genetically modified strains of microorganisms to convert the toxic xenobiotics to less toxic forms. Based on the site of bioremediation, the bioremediation strategies are categorized as in situ or ex situ (Fig. 6.1). In situ is performed at the site of contamination, whereas in ex situ bioremediation, the

J. S. Eswari et al., *Thermophiles for Biotech Industry*,
https://doi.org/10.1007/978-981-32-9919-1_6

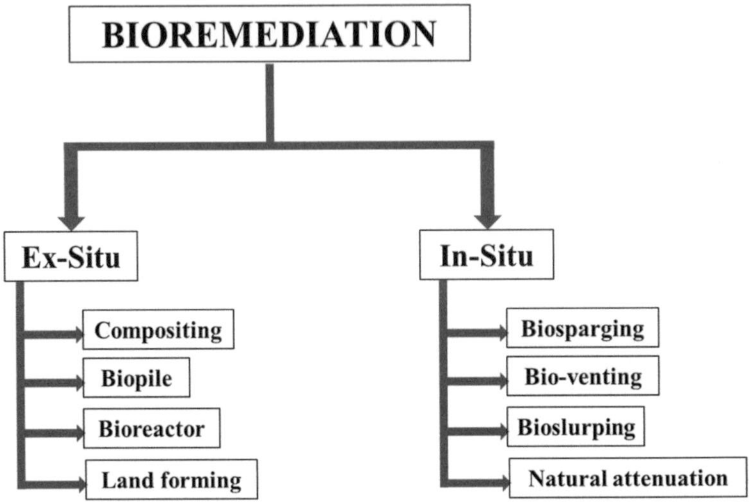

Fig. 6.1 Types of bioremediation strategies based on site of action

contaminated material is moved somewhere else for its treatment. There are many factors which affect the biodegradation of pollutants by microorganisms. They are the total nutrient availability, temperature, pH, and moisture level in the soil (Shah 2014).

As diverse microorganisms are involved in bioremediation to degrade the toxicity caused by environmental pollution, the interaction of microorganisms with the polluted environment becomes a significant factor to determine the efficacy of bioremediation process. Among various microbial species, few of them involved in bioremediation are *Pseudomonas aeruginosa*, *P. putida*, *Dechloromonas aromatica*, genetically engineered *Deinococcus radiodurans*, *Methylibium petroleiphilum*, *Alcanivorax borkumensis* as the bacterial species, and *Phanerochaete chrysosporium* as the fungal species. The domain archaea play an important role in bioremediation as they can survive in extreme conditions. As many industrial waste streams are hot, archaea, which are thermophilic or acidic by nature, offer suitable advantage over mesophilic microorganisms in the degradation of pollutants.

Thermophilic microorganisms have the ability to survive under high and stressful conditions. They have various strategies to deal with high concentrations of heavy metals and hence are used in bioremediation of heavy metals from environment at high temperatures. The unique structure of cell wall of thermophilic bacteria and their enzymatic and metabolic properties play an important role in interaction of thermophiles with metal (Fig. 6.2). Biosorption is a common approach for bioremediation wherein microbes are used to transform metal through oxidation/reduction reactions. This transformation can be in the form of changing the chemical form of metal (speciation), distribution or altering the toxicity of the metal. Thermophilic bacteria have a lot of applications in metal bioremediation and help in

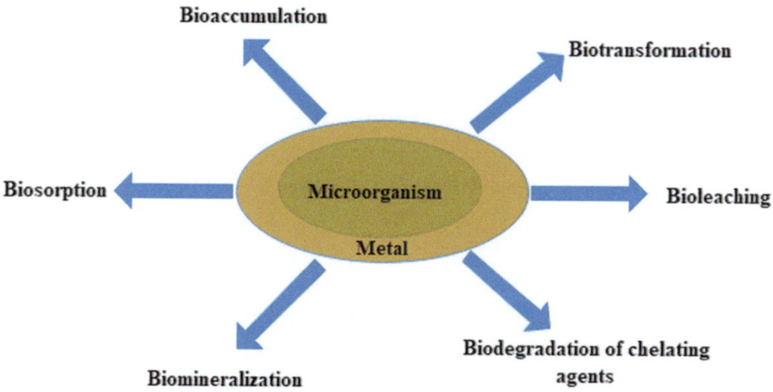

Fig. 6.2 Interaction of microorganisms with metal for bioremediation (Ahemad 2012)

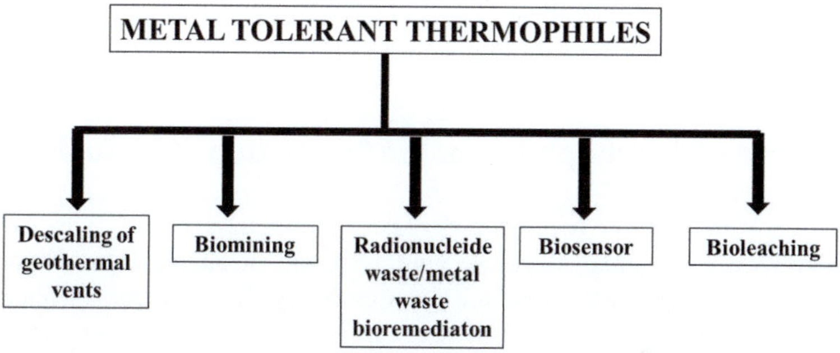

Fig. 6.3 Applications of metal-tolerant thermophiles in bioremediation (Ranawat and Rawat 2018)

solubilizing metals by oxidizing iron or sulfur. These bacteria can reduce metals like cobalt, chromium, gold, manganese, mercury, molybdenum and uranium (Sar et al. 2013). Few of the applications of metal-tolerant thermophiles are shown in Fig. 6.3. Apart from their role in biotransformation, thermophilic microorganisms are also used in microbial enhanced oil recovery processes.

Thermophiles have also been used for degradation of crude oil and petroleum contaminants (Mehta et al. 2016) along with the degradation of hydrocarbons. Thermophiles have shown to degrade aromatic, aliphatic, nitrated, and halogenated hydrocarbons. The solubility and bioavailability of hydrocarbons are increased at higher temperatures, and hence thermophiles are preferred for degradation of hydrocarbons. They also have an important role in bioremediation of oil-polluted desert soil (Margesin and Schinner 2001). Thermophilic bacterial strains isolated from local hot springs of United Arab Emirates exhibited maximum growth rate at 60–80 °C. The bacteria, belonging to genus *Bacillus*, demonstrated in situ and ex situ remediation of petroleum contaminants along with enhanced growth rate in the presence of crude oil. The biodegradation was enhanced due to the lowering of oil-water interfacial tension and formation of microemulsions (Al-Maghrabi et al. 1999). Thermophilic bacteria isolated from volcanic island were explored for the degradation of long-chain crude oil. The isolates were found to be belonging to *Bacillus* and *Geobacillus* species. These organisms were grown in the presence of crude oil as the sole carbon source and were capable of degrading 46.64–87.68% of the crude oil. This shows that *Bacillus* and *Geobacillus* species can be used for efficient removal of oil from polluted soil (Meintanis et al. 2006).

In a recent study, a biosurfactant from a marine bacterium *B. megaterium* and a biopolymer from thermophilic *B. licheniformis* were used together in oil recovery. The biosurfactant conditioned with Ca^{2+} ions was used to increase efficiency of oil recovery, and the biopolymer was used as flooding agent to increase the viscosity. Optimization techniques were exploited to determine the best conditions (concentration of lipopeptides and Ca^{2+} and pH of the solution) for oil recovery. The concentration of biosurfactant at 1.1 g/l and biopolymer at 3 g/l led to 45% of the total oil recovery showing the combined effect of biosurfactant and biopolymer in solubilization and mobilization of oil from soil (Dhanarajan et al. 2017).

2 Role of Thermophiles for the Production of Biofuels

Biofuel is any combustible fuel that is derived from biomass, namely, algae, microorganisms, plant, or animal wastes. In contrast to the ancient biological hydrocarbons, the raw materials of biofuel are from currently living microbial, plant, and animal matter. Biofuel is a source of renewable energy as unlike fossil fuels it is produced from readily available feedstock. As they are produced from cheap raw materials, they can be considered as best alternatives to petroleum-based fuels along with the reduction in greenhouse emissions. Biofuels can be categorized as liquid fuels (biodiesel and ethanol) and solid fuels (wood pellets). Based on the type of raw materials used for production, biofuels can be classified into first-generation, second-generation, and third-generation biofuels (Fig. 6.4). In first-generation biofuels, food crops are used as feedstock. For example, feedstock rich in sugar such as corn, maize, and sugarcane are used for the production of bioethanol, whereas vegetable oils, namely, soybean and canola, are used for biodiesel production. As first-generation biofuel production requires food crop, hence it is not the preferred method for biofuel production. Second-generation biofuel requires cellulosic material as feedstock. The disadvantages of this feedstock are that high cost and energy are required for pretreatment of lignocellulosic biomass through fermentation. The third generation of biofuels utilizes algal lipids for biofuel production. The biofuel production from wastes such as spent cooking oil, garbage, and animal fats is termed as advanced biofuels (Naik et al. 2010).

Due to the ever-increasing demand of energy sources and ever-decreasing reserves of fossil fuels, there is a need to search for alternative fuel sources. Biofuels are classified as biobutanol, biodiesel, bioethanol, and biogas. The microbial strains

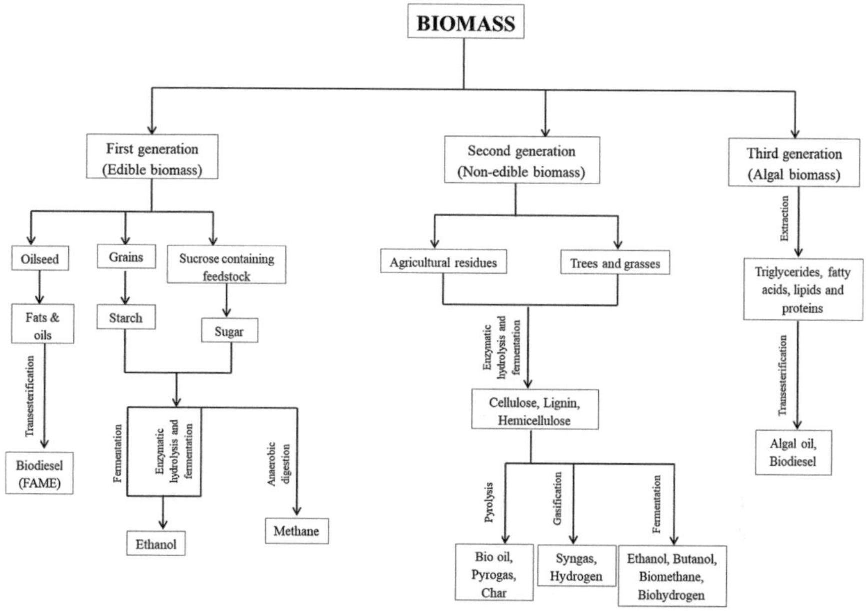

Fig. 6.4 Classification of biofuels based on the type of feedstock used (Khatiwada 2013; Nanda et al. 2018)

for the production of bioethanol should yield high amounts of ethanol, should be tolerant to ethanol, and should produce fewer side effects. Species of *Clostridia*, *Thermoanaerobacterium*, and *Geobacillus* are some of the few thermophilic species of bacteria used for the production of bioethanol (Barnard et al. 2010). Consolidated bioprocessing is a process for the conversion of lignocellulosic biomass to bioethanol using thermophilic microorganisms without the requirement of an additional step of cellulose hydrolysis. The process requires co-fermentation of a thermophilic cellulolytic bacterium (*Clostridium thermocellum*) with thermophilic bacterium to produce ethanol at 55–60 °C. Svetlitchnyi et al. isolated seven strains of thermophilic cellulolytic bacteria belonging to *Caldicellulosiruptor* and eight strains of *Thermoanaerobacter* capable of degrading xylan. These organisms had an optimum growth temperature of 72 °C with high lignocellulose degradation and bioethanol production activities. Strains of *Caldicellulosiruptor* degraded up to 89–97% of cellulose and hemicellulose with the production of 72 mM of ethanol from the degraded cellulose, whereas co-cultures of *Caldicellulosiruptor* with *Thermoanaerobacter* increased the ethanol yield from 2 to 8.2-fold. More specifically, the co-cultures of *Caldicellulosiruptor* DIB 087C and *Thermoanaerobacter* DIB 097X were able to convert 34.8 mol% of the total organic products to bioethanol. This depicts the importance of using a combination of different bacterial species for high ethanol production (Svetlitchnyi et al. 2013).

3 Importance of Thermophiles in Biorefinery

Biorefinery is a facility that converts biomass to heat, power, fuels, and value-added chemicals. In this process, different biochemicals are produced from the different components of biomass and its intermediates. Hence, biorefinery utilizes the biomass feedstock to obtain a wide range of products with maximum value. These products can be chemicals, materials for food and feed, or bioenergy. Biorefinery considers different technologies to isolate building blocks (biomolecules, such as, carbohydrates, lipids, and proteins) from biomass (grasses, agricultural residues, wood, etc.) for their conversion to value-added biochemical products and biofuels. The concept of biorefinery is analogous to petroleum industry wherein chemical fuels and products are produced from petroleum. Biorefineries produce either high-value low volume or low-value high volume products from low-value biomass. The advantage of using biorefinery is the increased volume of fuel and power to meet the energy needs, reduced greenhouse gas production, and profit from the production of high-value products. The important and much-awaited future aspect of biorefinery will be to produce chemicals that are conventionally produced from petroleum (Cherubini 2010). As per IEA Bioenergy Task 42, "Biorefining is the sustainable processing of biomass into a spectrum of marketable products and energy" (de Jong et al. 2012).

The high amount of proteins, carbohydrates, and lipids in microalgae can be used as suitable feedstock for various other products. Biorefinery plays an important role in utilizing all the products of microalgae. The main challenge of this technique is the separation of different components without damaging them. The first step is the disruption of cells to release the products in the culture medium (Fig. 6.5). After the release of products, their extraction and separation are performed for purification. These techniques must be suitable and in accordance with the algal species. The organelles of algal cells, namely, mitochondria, lysosome, and endoplasmic reticulum, have different components that can be exploited for the extraction of different products. A general extraction process first disrupts the cell for the release of lipids, carbohydrates, and proteins, and then the larger cellular organelles can be disrupted to obtain their specific biomolecules (Eppink et al. 2013). The microalgal lipids can be used as potential substrate for biodiesel production, whereas carbohydrates are utilized as carbon source in fermentation industries for the production of ethanol and other platform chemicals instead of conventional carbohydrates and lignocellulosic biomass. The pigments and proteins in microalgal biomass can be used in pharmaceutical industries and food and feed market and for the production of bulk chemicals, while the long-chain fatty acids are used as health supplements, as source of biofuels, as edible oils in food, or as a building block in biochemical industries. The carbohydrates, lipids, and proteins of microalgae can also be utilized for the production of biohydrogen and methane by anaerobic digestion. Microalgae can also yield bio-oil, syngas, biobutanol, and fuel by different biochemical, chemical, and thermochemical processes. Thus microalgae are ideal for the production of biochemical compounds, cosmetics, pharmaceuticals, nutrition, and animal feed (Vanthoor-Koopmans et al. 2013; Zhu 2015; Chew et al. 2017).

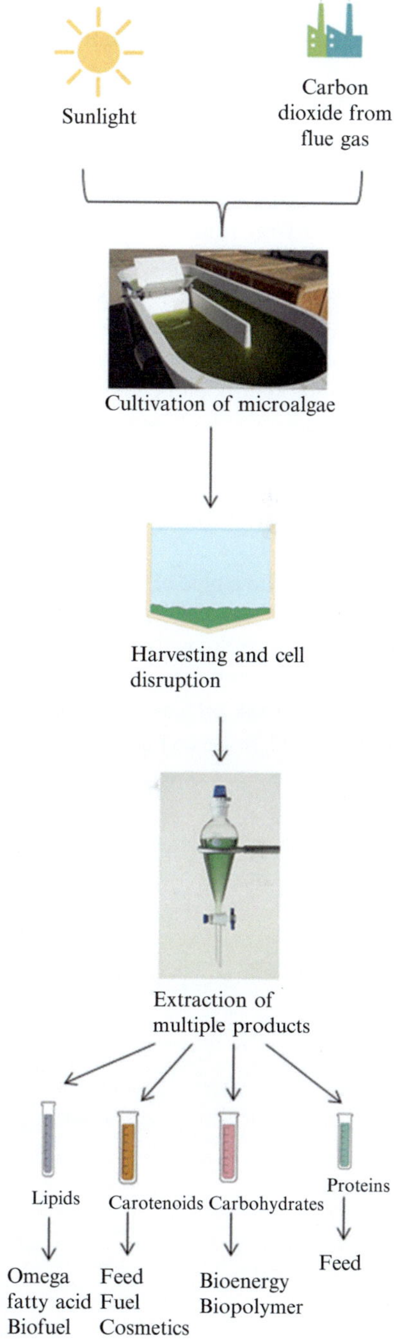

Fig. 6.5 Microalgae as biorefinery (Minhas et al. 2016)

Temperature has a strong influence on the growth of microalgae. Each microalgal species has an optimal temperature requirement, and variation in that temperature affects the cellular enzymatic reaction of that organism. At high temperatures, the essential proteins are denatured, and the cellular physiology is inhibited due to the denaturation of photosystem II. The CO_2 fixation, at higher temperatures, is also affected by the increase in photorespiration and the decrease in carbon dioxide to oxygen solubility ratio. In the bioremediation process, microalgae are employed to sequester CO_2 from flue gas. The flue gas from industrial plants generally has a higher temperature, ranging from 70 to 120 °C, and hence thermophilic species of microalgae are preferred. Also, most of the microalgal cultivation is performed outdoors with variable temperature. The countries near the equator have higher temperatures, whereas the countries far away from the equator have lower temperatures. Therefore, thermophilic algae become an important alternative for CO_2 sequestration processes at high temperatures (Jacob-Lopes et al. 2015; Kumar et al. 2015).

4 Significance of Thermophiles in Detergent Industries

Detergents are a type of surface-active molecules that remove foreign material from stained surfaces while retaining them in suspension. Many enzymes, such as proteases, lipases, cellulase, and amylase, are used in detergent industries to break down proteins, fats, cellulose, and starch for removal of stains (Fig. 6.6). Among the different enzymes, proteases are a major group of industrial enzymes. They have been widely used in detergents due to the availability of a number of detergent-stable proteases (Kumar et al. 2008).

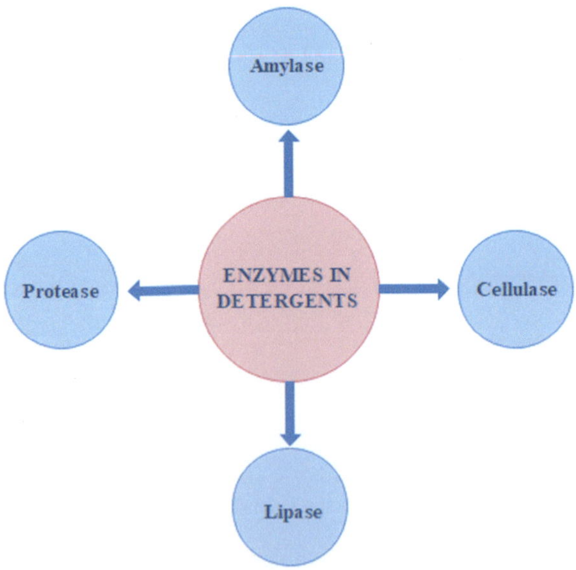

Fig. 6.6 Various enzymes used in detergent industry

Approximately 25% of the proteases produced worldwide are used in detergent industries. These enzymes hydrolyze large insoluble proteins that are removed from the surface of fabric (Hmidet et al. 2009). Proteases from thermophilic organisms are structurally stable at high temperatures as they are produced from organisms that are adapted to high temperatures. The advantage of thermophilic proteases is that they can be used in processes where higher temperature is required to carry out the process. This also results in faster rates of reaction (Nascimento and Martins 2004).

An alkalophilic strain of *Bacillus licheniformis* NH1 has shown to produce five extracellular proteases and an amylase. The optimum conditions of the amylase were 90 °C and pH of 6.5, whereas those of proteases were 70 °C and 10. The enzymes showed stability toward oxidizing agents. The crude enzymes from NH1 were effective in removing stains of chocolate, blood, and sauce. This shows the potential application of enzymes from *B. licheniformis* NH1 in detergent industries (Hmidet et al. 2009). The protease enzyme from another species of *Bacillus*, *Bacillus* RV.B2.90, was purified and tested for its cleaning abilities. The enzyme could degrade pigment stains of beetroot, carrot, green leaves, tea and coffee, blood clot, and coagulated eggs. It also exhibited good compatibility with local commercial detergents. The maximum production of protease from this bacterium has been shown to be at 70 °C. The yield at this temperature was found to be 1984 U/ml. Protease from *Bacillus* RV.B2.90 with thermostability and alkalophilic properties has a great potential in detergent industry (Vijayalakshmi et al. 2011). *B. subtilis* WIFD5, isolated from milk powder sample, also produces thermophilic protease. The maximum activity of enzyme was produced at 55 °C when pH is 9. The activity of enzyme was enhanced in the presence of calcium and magnesium ions. This suggests that these ions are required as the cofactors for this enzyme. This enzyme demonstrated excellent blood stain removal ability when added to detergent as compared to detergent alone proving it to be a potential detergent additive (Sharma and Aruna 2015).

Cellulase is another enzyme which is used in detergent industry as softening, color-reviving, and anti-pilling agent. It is generally used for softening cotton and finishing denim in textile industry. It modifies cellulose fibrils to improve color and remove dirt from cotton fabrics (Kuhad et al. 2011). It is commercially produced by fungi. Bacteria have also been exploited to produce this enzyme as they are stable at extreme conditions and have high growth rate. Fungal cellulases hydrolyze crystalline cellulose, whereas bacterial cellulose can hydrolyze only synthetic carboxymethylcellulose and not crystalline cellulose, and hence fungal cellulose is more preferred in detergent industries. The disadvantage with fungal enzymes is that they are acidic and neutral enzymes and hence cannot hydrolyze cellulose in alkaline region.

Taking the lignocellulosic materials such as sugarcane bagasse and corn steep liquor as substrates, Ladeira et al. produced cellulase by thermophilic bacterium, *Bacillus* sp. SMIA-2. The bacterium produced cellulase at 50 °C in liquid cultures with lignocellulosic waste as substrate. The organism was capable of producing two types of cellulase, namely, avicelase and CMCase. The avicelase was produced at

Table 6.1 Microbial enzymes used in detergent industries (Kumar et al. 2008)

Microorganisms	Trade name	Producer
Bacillus sp.	Protosol	Advance Biochemicals, India
Bacillus sp.	Durazyme	Novozymes, Denmark
Bacillus sp.	Wuxi	Wuxi Synder Bioproducts, China
B. alcalophilus	KAP	Kao Corporation, Japan
B. alcalophilus PB92	Properase	Genencor International, USA
B. amyloliquefaciens	FNA	Genencor International, USA
B. amyloliquefaciens	Savinase	Novozymes, Denmark
B. clausii	Everlase	Novozymes, Denmark
B. clausii	Kannase	Novozymes, Denmark
B. lentus	BLAP S	Henkel, Germany
B. lentus	BLAP X	Henkel, Germany
B. lentus	Purafect	Genencor International, USA
B. licheniformis	Alcalase	Novozymes, Denmark
B. licheniformis, B. halodurans	Esperase	Novozymes, Denmark
B. licheniformis	Godo-Bap	Godo Shusei, Japan
B. subtilis	Bioprase	Nagase Biochemicals, Japan

120 h post incubation and pH 7.5 with a maximum activity of 0.83 U/ml. The maximum activity of CMCase was 0.29 U/ml at pH 8.0 and after 168 h of incubation. The optimum temperature of these enzymes was 70 °C, and these enzymes were stable at 60 °C for 1 h. The cellulase was resistant to degradation by protease produced by the same organism and was stable even in the presence of sodium dodecyl sulfate. These properties suggest that thermophilic *Bacillus* sp. SMIA 2 is a potential source of heat-stable cellulase with applications in detergent industry (Ladeira et al. 2015). Few of the examples of commercially used enzymes with their producer industries and microorganisms are mentioned in Table 6.1.

5 Potential of Thermophiles in Food and Beverage Industries

Biotechnology has a huge role in food industries to improve yield, taste, nutritive values, and shelf life. Processing of nonedible and perishable raw materials to edible food with longer shelf life using microorganisms is termed as fermentation. A normal diet across the globe consists of almost 30% of fermented food. Biotechnology also helps in preventing the contamination of food by the production of antimicrobial compounds that kill unwanted microbes. It also helps in the degradation of toxins and some anti-nutritional elements present in food. Microbial enzymes have also been used in food industries from a very long time for enhancing the flavor and shelf-life of food. This is achieved by the use of improved microbial strains and microbial derivatives (Fig. 6.7).

Starch processing industry is one of the largest consumers of enzymes. Enzymatic hydrolysis of starch yields monosaccharide (glucose), disaccharide (maltose), and oligosaccharides which can be used for the synthesis of alcohols, acids, amino

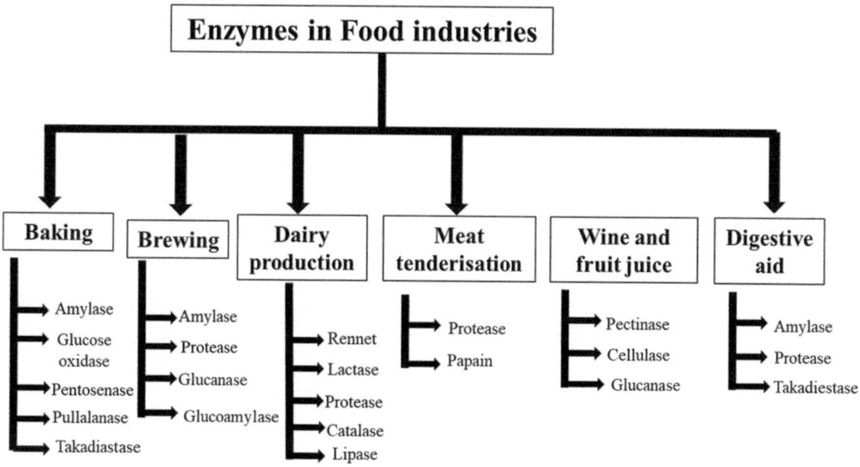

Fig. 6.7 Different microbial enzymes used in food industries

acids, etc. Hydrolysis of starch through liquefaction and saccharification is performed at high temperatures (above 60–70 °C) (Bentley and Williams 1996). As enzymes from hyperthermophiles work best at 80–110 °C, they are preferred for liquefaction of starch (Niehaus et al. 1999). This requires the need of new hyperthermophilic enzymes to be used in industries. Hence, thermostable α-amylases from hyperthermophiles, such as *Clostridium thermocellum* SS8 and *Thermoanaerobacterium thermosulfurigenes*, have been exploited for saccharification process (Kitamoto et al. 1988; Swamy et al. 1994). Mutant *Escherichia coli* and *Pichia pastoris* have been created by inserting heat-stable α-amylases from *B. licheniformis* (Du et al. 2006; Sarmiento et al. 2015). Amylases are used in brewing and baking to improve freshness of bread and produce cakes, fruit juices, and digestive aids.

Amylopullulanases are also starch processing enzymes which yield maltose and maltotriose syrups. They are bifunctional enzymes and can catalyze debranching and liquefying reactions independent of calcium ions and hence are better than α-amylases and β-amylases. Thermophilic amylopullulanase have been cloned in *E. coli* and *B. subtilis* for their increased expression (Nisha and Satyanarayana 2013).

Glucose isomerase isomerizes glucose to fructose for the formation of fructose syrups. This reaction is performed at temperatures of 60–90 °C. Glucose or xylose isomerases are isolated from *Streptomyces murinus*, *Thermotoga maritima*, *Thermotoga neapolitana*, *Thermus aquaticus*, and *Thermus thermophilus*. *Bacillus* sp. and *Fervidobacterium gondwanense* were used for the production of thermophilic glucose isomerases. These enzymes had optimal activity at 70 °C and hence can be used for industrial applications requiring high temperature (Sukumar et al. 2013).

Probiotics are microorganisms that help to improve microbial flora of stomach and hence provide health benefits to the consumers. The criteria for considering a microorganism as a probiotic are that the microbial strain should be safe and should preferably originate from animal microflora. They should be resistant to the in vivo conditions, have antimicrobial properties against pathogens, and should adhere to the intestine. It should also be resistant to the harsh conditions of processing and storage. The spray drying of the probiotics leads to the death of mesophiles due to the damage in their DNA, cell wall, and cell membrane. This challenge can be overcome by the use of thermophilic microbial strains as probiotics. *Lactobacillus acidophilus*, *L. helveticus*, and *L. delbrueckii* ssp. *bulgaricus* and ssp. *lactis* have been identified as thermophilic lactic acid bacteria in dairy products (Andrighetto et al. 1998). Other thermotolerant species of lactic acid bacteria isolated from chicken, namely, *L. reuteri*, *L. thermotolerans*, *L. aminata*, and *L. paraplantarum*, have also been used as probiotic cultures (Nitisinprasert et al. 2000; Dauphin et al. 2005; Kosin and Rakshit 2006).

References

Ahemad M (2012) Implications of bacterial resistance against heavy metals in bioremediation: a review. J Inst Integr Omics Appl Biotechnol (IIOAB) 3(3):39–46

Al-Maghrabi IM et al (1999) Use of thermophilic bacteria for bioremediation of petroleum contaminants. Energy Sources 21(1–2):17–29

Andrighetto C et al (1998) Molecular identification and cluster analysis of homofermentative thermophilic lactobacilli isolated from dairy products. Res Microbiol 149(9):631–643

Barnard D et al (2010) Extremophiles in biofuel synthesis. Environ Technol 31(8–9):871–888

Bentley I Williams E (1996) Starch conversion. Ind Enzymol: 341–357

Cherubini F (2010) The biorefinery concept: using biomass instead of oil for producing energy and chemicals. Energy Convers Manag 51(7):1412–1421

Chew KW et al (2017) Microalgae biorefinery: high value products perspectives. Bioresour Technol 229:53–62

Dauphin RD et al (2005) Polyphasic identification of a new thermotolerant species of lactic acid bacteria isolated from chicken faeces. Afr J Biotechnol 4(5):409–421

de Jong E et al (2012) Bio-based chemicals value added products from biorefineries. IEA Bioenergy, Task42 Biorefinery

Dhanarajan G et al (2017) Biosurfactant-biopolymer driven microbial enhanced oil recovery (MEOR) and its optimization by an ANN-GA hybrid technique. J Biotechnol 256:46–56

Du B et al (2006) Expression of a thermostable a-amylase mutant into Escherichia coli and Pichia pastoris. Wei Sheng Wu Xue Bao = Acta Microbiol Sin 46(5):827–830

Eppink M et al (2013) Biorefinery of microalgae: production of high value products, bulk chemicals and biofuels. In: Symposium Biorefinery for Food, Fuel and Materials 2013

Hmidet N et al (2009) Alkaline proteases and thermostable α-amylase co-produced by Bacillus licheniformis NH1: characterization and potential application as detergent additive. Biochem Eng J 47(1–3):71–79

Jacob-Lopes E et al (2015) Microalgal biorefineries. In: Biomass production and uses. IntechOpen

Khatiwada D (2013) Assessing the sustainability of bioethanol production in different development contexts: a systems approach. KTH Royal Institute of Technology

Kitamoto N et al (1988) Cloning and sequencing of the gene encoding thermophilic beta-amylase of Clostridium thermosulfurogenes. J Bacteriol 170(12):5848–5854

Kosin B, Rakshit SK (2006) Microbial and processing criteria for production of probiotics: a review. Food Technol Biotechnol 44(3):371–379

Kuhad RC et al (2011) Microbial cellulases and their industrial applications. Enzym Res 2011:280696

Kumar D et al (2008) Microbial proteases and application as laundry detergent additive. Res J Microbiol 3(12):661–672

Kumar K et al (2015) CO 2 sequestration through algal biomass production. In: Algal biorefinery: an integrated approach. Springer, p 35–57

Ladeira SA et al (2015) Cellulase production by thermophilic Bacillus sp. SMIA-2 and its detergent compatibility. Electron J Biotechnol 18(2):110–115

Margesin R, Schinner F (2001) Biodegradation and bioremediation of hydrocarbons in extreme environments. Appl Microbiol Biotechnol 56(5–6):650–663

Mehta R et al (2016) Insight into thermophiles and their wide-spectrum applications. 3 Biotech 6(1):81

Meintanis C et al (2006) Biodegradation of crude oil by thermophilic bacteria isolated from a volcano island. Biodegradation 17(2):3–9

Minhas AK et al (2016) A review on the assessment of stress conditions for simultaneous production of microalgal lipids and carotenoids. Front Microbiol 7:546

Naik SN et al (2010) Production of first and second generation biofuels: a comprehensive review. Renew Sust Energ Rev 14(2):578–597

Nanda S et al (2018) A broad introduction to first-, second-, and third-generation biofuels. Recent advancements in biofuels and bioenergy utilization. Springer, pp 1–25

Nascimento WCA d, Martins MLL (2004) Production and properties of an extracellular protease from thermophilic Bacillus sp. Braz J Microbiol 35(1–2):91–96

Niehaus F et al (1999) Extremophiles as a source of novel enzymes for industrial application. Appl Microbiol Biotechnol 51(6):711–729

Nisha M, Satyanarayana T (2013) Recombinant bacterial amylopullulanases: developments and perspectives. Bioengineered 4(6):388–400

Nitisinprasert S et al (2000) Screening and identification of effective thermotolerant lactic acid bacteria producing antimicrobial activity against Escherichia coli and Salmonella sp. resistant to antibiotics. Kasetsart J (Nat Sci) 34:387–400

Ranawat P, Rawat S (2018) Metal-tolerant thermophiles: metals as electron donors and acceptors, toxicity, tolerance and industrial applications. Environ Sci Pollut Res 25(5):4105–4133

Sar P et al (2013) Metal bioremediation by thermophilic microorganisms. In: Thermophilic microbes in environmental and industrial biotechnology. Springer, pp 171–201

Sarmiento F et al (2015) Cold and hot extremozymes: industrial relevance and current trends. Front Bioeng Biotechnol 3:148

Shah MP (2014) Environmental bioremediation: a low cost nature's natural biotechnology for environmental clean-up. J Pet Environ Biotechnol 5(4):1

Sharma S, Aruna K (2015) Optimization of protease activity of thermophilic Bacillus subtilis WIFD5 isolated from milk powder. Biomed Pharmacol J 5(1):57–64

Sukumar M et al (2013) Production and partial characterization of extracellular glucose isomerase using thermophilic Bacillus sp. isolated from agricultural land. Biocatal Agric Biotechnol 2(1):45–49

Svetlitchnyi VA et al (2013) Single-step ethanol production from lignocellulose using novel extremely thermophilic bacteria. Biotechnol Biofuels 6(1):31

Swamy M et al (1994) β-Amylase from Clostridium thermocellum SS8-a thermophilic, anaerobic, cellulolytic bacterium. Lett Appl Microbiol 18(6):301–304

Vanthoor-Koopmans M et al (2013) Biorefinery of microalgae for food and fuel. Bioresour Technol 135:142–149

Vijayalakshmi S et al (2011) Screening of alkalophilic thermophilic protease isolated from Bacillus RV. B2. 90 for industrial applications. Res Biotechnol 2(3):32–41

Zhu L (2015) Biorefinery as a promising approach to promote microalgae industry: an innovative framework. Renew Sust Energ Rev 41:1376–1384

MIX
Papier aus verantwortungsvollen Quellen
Paper from responsible sources
FSC® C105338

If you have any concerns about our products,
you can contact us on
ProductSafety@springernature.com

In case Publisher is established outside the EU,
the EU authorized representative is:
Springer Nature Customer Service Center GmbH
Europaplatz 3, 69115 Heidelberg, Germany

Printed by Libri Plureos GmbH
in Hamburg, Germany